KU-296-253

Contents

Preface

Digital video was conceptually made possible by the invention of pulse code modulation, but the technology required to implement it did not then exist. Experiments were made with digital video from the late 1950s onwards, but the cost remained prohibitive for commercial applications until sufficient advances in microelectronics and high-density recording had been made. Early digital broadcast and production equipment was large and expensive and was handled by specialists, but now the technology has advanced to the point where traditional analog equipment is being replaced by digital equipment on economic grounds. Soon, digital video equipment will become the norm in broadcast and production applications. Consequently, increasing numbers of people are being exposed to digital video equipment as it enters the mainstream. Digital video equipment uses techniques which have no parallels in analog video and it requires new skills. This book is designed to introduce in an understandable manner the theory behind digital video and to illustrate it with current practice. Whilst some aspects of the subject are complex, and are generally treated mathematically, such treatments are not appropriate for a mainstream book and have been replaced wherever possible by plain English supported by diagrams. Many of the explanations here have evolved during the many training courses and lectures I have given on the subject to students from all technical backgrounds. No great existing knowledge of digital techniques is assumed, as every topic here evolves from first principles. As an affordable introductory book, this volume concentrates on essentials and basics. Readers requiring an in-depth treatment of the subject are recommended to the companion works *The Art of Digital Video* and *The Digital Video Tape Recorder*.

John Watkinson
Burghfield Common 1994

Chapter 1

Introducing digital video

1.1 Analog video

The first methods used in television transmission and recording were understandably analog, and the signal formats were essentially determined by the requirements of the cathode ray tube as a display. Study of every shortcoming in the analog production process has led to the development of some measures to reduce them. However, analog production equipment has become a mature technology, and has almost reached limits determined by the laws of physics.

As there is now another video technology, known as digital, it is appropriate to begin by contrasting the two technologies. Comparison is only possible in areas where the two technologies overlap. The real strength of digital video lies in areas where no comparison is possible because such processes are simply impossible in the analog domain.

In an analog system, information is conveyed by some infinite variation of a continuous parameter such as the voltage on a wire or the strength or frequency of flux on a tape. In a recorder, distance along the medium is a further, continuous, analog of time. It does not matter at what point a recording is examined along its length – a value will be found for the recorded signal. That value can itself change with infinite resolution within the physical limits of the system.

Those characteristics are the main weakness of analog signals. Within the allowable bandwidth, *any* waveform is valid. If the speed of the medium is not constant, one valid waveform is changed into another valid waveform; a timebase error cannot be detected in an analog system. In addition, a voltage error simply changes one valid voltage into another; noise cannot be detected in an analog system. We might suspect noise, but how is one to know what proportion of the received signal is noise and what is the original? If the transfer function of a system is not linear, distortion results, but the distorted waveforms are still valid; an analog system cannot detect distortion.

It is a characteristic of analog systems that degradations cannot be separated from the original signal, so nothing can be done about them. At the end of a system a signal carries the sum of all degradations introduced at each stage through which it passed. This sets a limit to the number of stages through which a signal can be passed before it is useless.

1.2 The characteristics of video signals

Video signals are electrical waveforms which allow moving pictures to be conveyed from one place to another. Observing the real world with the human eye results in a two-dimensional image on the retina. This image changes with time and so the basic information is three-dimensional. With two eyes a stereoscopic view can be obtained and stereoscopic television is possible with suitable equipment. However this is restricted to specialist applications and has not been exploited in broadcasting.

An electrical waveform is two-dimensional in that it carries a voltage changing with respect to time. In order to convey three-dimensional picture information down a two-dimensional cable it is necessary to resort to scanning. Instead of attempting to convey the brightness of all parts of a picture at once, scanning conveys the brightness of a single point which moves with time.

The cathode ray tube is not a linear device, but produces light following a power law with an index of about 2.2. This non-linear characteristic is known as *gamma*. In practice an inverse law is applied in the camera so that no non-linear circuitry is needed in the television receiver. Thus recording and processing is performed on compressed signals. Perhaps surprisingly, the presence of the non-linearity makes little practical difference.

In television systems the scan consists of rapid horizontal sweeps combined with a slower vertical sweep so that the image is scanned in lines. At the end of each vertical sweep, or frame, the process recommences. Computer monitors scan in this way, but in most broadcast systems the scanning process is 2:1 interlaced. In an interlaced system the vertical sweep speed is doubled so that spaces open up between the scanned lines. The vertical scan, or field, takes half as long and contains half as many lines. In the second field the areas which were missed in the first field are scanned. Figure 1.1(a) shows that this is readily achieved by having an odd number of lines, e.g. 525 or 625, in the frame so that the first field begins with a whole line and ends half-way along a line and the second field begins half-way through a line and ends on a whole line. The lines of the two fields then automatically mesh vertically.

The scanning process converts resolution on the image into the frequency domain. The higher the resolution of the image, the more lines are necessary to resolve the vertical detail. The line rate is increased along with the number of cycles of modulation which need to be carried in each line. If the frame rate remains constant, the bandwidth goes up as the square of the resolution. A 625 line system has a bandwidth of nearly 6 MHz, whereas a 1250 line HDTV system having twice the resolution needs about 30 MHz. The bandwidth required is more than four times higher because HDTV uses a 16:9 aspect ratio instead of 4:3, and so the lines are longer.

In interlaced systems, the field rate is intended to determine the flicker frequency, whereas the frame rate determines the bandwidth needed, which is thus halved along with the information rate. Information theory tells us that halving the information rate must reduce quality, and so the saving in bandwidth is accompanied by a variety of effects. Figure 1.1(b) shows the spatial/temporal sampling points in a 2:1 interlaced system. If an object has a sharp horizontal edge, it will be present in one field but not in the next. The refresh rate of the edge will be reduced to frame rate, 25 Hz or 30 Hz, and becomes visible as twitter.

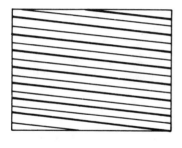

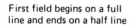

First field begins on a full
line and ends on a half line

Second field begins on a half
line and ends on a full line

There must be an odd number of lines in a frame

Figure 1.1(a) 2:1 interlace.

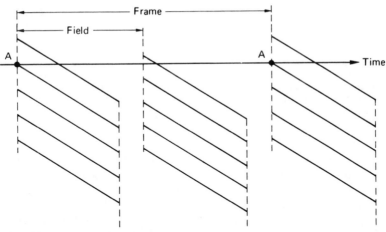

Figure 1.1(b) In an interlaced system, a given point A is only refreshed at frame rate, causing twitter on fine vertical detail.

Figure 1.2 shows some of the basic types of analog colour video. Each of these types can, of course, exist in a variety of line standards. Since practical colour cameras generally have three seperate sensors, one for each primary colour, an *RGB* system will exist at some stage in the internal workings of the camera, even if it does not emerge in that form. *RGB* consists of three parallel signals each having the same spectrum, and is used where the highest accuracy is needed, often for production of still pictures. Examples of this are paint systems and in computer-aided design (CAD) displays. *RGB* is seldom used for real-time video

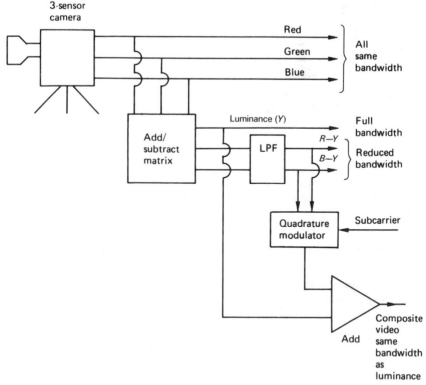

Figure 1.2 The major types of analog video. Red, green and blue signals emerge from the camera sensors, needing full bandwidth. If a luminance signal is obtained by a weighted sum of R, G and B, it will need full bandwidth, but the colour difference signals $R-Y$ and $B-Y$ need less bandwidth. Combining $R-Y$ and $B-Y$ into a subcarrier modulation scheme allows colour transmission in the same bandwidth as monochrome.

recording because of the high cost. As the red, green and blue signals directly represent part of the image, this approach is known as component video.

Some saving of bandwidth can be obtained by using colour difference working. The human eye relies on brightness to convey detail, and much less resolution is needed in the colour information. R, G and B are matrixed together to form a luminance (and monochrome-compatible) signal Y which has full bandwidth. The eye is not equally sensitive to the three primary colours, as can be seen in Figure 1.3(a), and so the luminance signal is a weighted sum.

The origin of the common colour bar test signal is shown in Figure 1.3(b). Binary RGB signals are produced, having one, two and four cycles per screen width. When these are added together, an eight-level luminance staircase results because of the unequal weighting. The matrix also produces two colour difference signals, $R - Y$ and $B - Y$, and these are often displayed simultaneously on a vectorscope as shown in Figure 1.3(c). Note that the white bar and the black bar are not colours and so the colour difference signal is zero in those bars, resulting in a central spot. There are six remaining colours, each of which results in a spot on the perimeter of the vectorscope display.

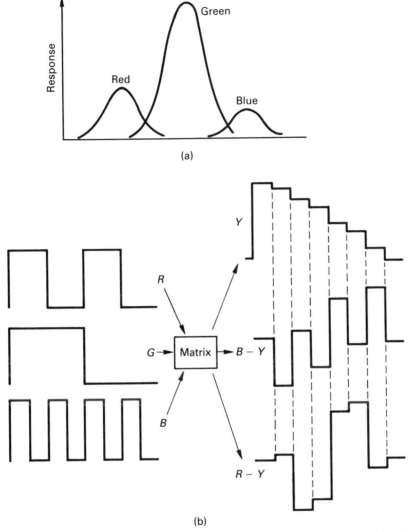

(a)

(b)

Figure 1.3 The response of the human eye to colour is not uniform as shown in (a). Colour bars originate as three square waves in *RGB*, but luminance obtained by weighted adding follows an irregular staircase shown in (b).

Colour difference signals do not need the same bandwidth as *Y*, because the eye's acuity does not extend to colour vision. One-half or one-quarter of the bandwidth will do depending on the application. Chroma keying is more accurate with wide colour difference bandwidth, and so production systems use one-half of the luminance bandwidth for the colour difference signals. The overall bandwidth is then two-thirds that needed by *RGB*. Analog colour difference recorders such as Betacam and M II record the luminance and colour difference signals separately. The D-1, D-5 and Digital Betacam formats record 525/60 or 625/50 colour difference signals digitally. In casual parlance, colour difference

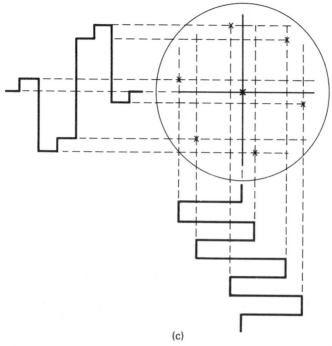

(c)

Figure 1.3 continued Colour difference signals can be shown two-dimensionally on a vectorscope as in (c).

formats are often called component formats to distinguish them from composite formats.

For monochrome-compatible colour television broadcast in a single channel, the NTSC, PAL and SECAM systems interleave into the spectrum of a monochrome signal a subcarrier which carries two colour difference signals of restricted bandwidth. The subcarrier is intended to be invisible on the screen of a monochrome television set. A subcarrier-based colour system is generally referred to as composite video, and the modulated subcarrier is called chroma.

PAL and SECAM share the same scanning standard of 625/50, but the methods of conveying the colour are totally incompatible. NTSC scans at 525/59.94, but the colour system has a few things in common with PAL. All three standards use 2:1 interlace. Analog composite recorders include the B and C formats, and the D-2 and D-3 formats record PAL or NTSC digitally.

1.3 What is digital video?

One of the vital concepts to grasp is that digital video is simply an alternative means of carrying a video waveform. An ideal digital video system has the same characteristics as an ideal analog system: both of them are totally transparent and reproduce the original applied waveform without error. One need only compare high-quality analog and digital equipment side by side with the same signals to realize how transparent modern equipment can be. Needless to say in the real world ideal conditions seldom prevail, so analog and digital equipment both fall

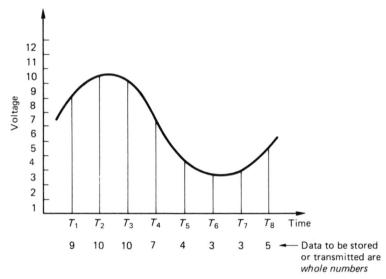

Figure 1.4 In pulse code modulation (PCM) the analog waveform is measured periodically at the sampling rate. The voltage (represented here by the height) of each sample is then described by a whole number. The whole numbers are stored or transmitted rather than the waveform itself.

short of the ideal. Digital equipment simply falls short of the ideal by a smaller distance than does analog and at lower cost, or, if the designer chooses, can have the same performance as analog at much lower cost.

Although there are a number of ways in which video waveforms can be represented digitally, there is one system, known as pulse code modulation (PCM) which is in virtually universal use.[1] Figure 1.4 shows how PCM works. Instead of being continuous, the time axis is represented in a discrete, or stepwise manner. The waveform is not carried by continuous representation, but by measurement at regular intervals. This process is called sampling and the frequency with which samples are taken is called the sampling rate or sampling frequency F_s. It should be stressed that sampling is an analog process. Each sample still varies infinitely as the original waveform did. Sampled analog devices are well known in video. The charge-coupled sensor in a television camera is one example. To complete the conversion to PCM, each sample is then represented to finite accuracy by a discrete number in a process known as quantizing.

In television systems the input image which falls on the camera sensor will be continuous in time, and continous in two spatial dimensions corresponding to the height and width of the sensor. In analog video systems, the time axis is sampled into frames, and the vertical axis is sampled into lines. Digital video simply adds a third sampling process along the lines.

There is a direct connection between the concept of temporal sampling, where the input changes with respect to time at some frequency and is sampled at some other frequency, and spatial sampling, where an image changes a given number of times per unit distance and is sampled at some other number of times per unit distance. The connection between the two is the process of scanning. Temporal

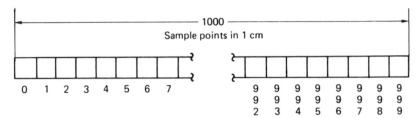

Figure 1.5 If the above spatial sampling arrangement of 1000 points per centimetre is scanned in 1 millisecond, the sampling rate will become 1 megahertz.

frequency can be obtained by multiplying spatial frequency by the speed of the scan. Figure 1.5 shows a hypothetical image sensor which has 1000 discrete sensors across a width of 1 centimetre. The spatial sampling rate of this sensor is thus 1000 per centimetre. If the sensors are measured sequentially during a scan which takes 1 millisecond to go across the 1 centimetre width, the result will be a temporal sampling rate of 1 MHz. In the spatial domain we refer to resolution, whereas in the time domain we refer to frequency response. The two are linked by scanning.

Whilst any sampling rate which is high enough could be used for video, it is common to make the sampling rate a whole multiple of the line rate. Samples are then taken in the same place on every line. If this is done, a monochrome digital image is a rectangular array of points at which the brightness is stored as a number. The points are known as picture cells, generally abbreviated to pixels, although sometimes the abbreviation is more savage and they are known as pels. As shown in Figure 1.6(a), the array will generally be arranged with an even spacing between pixels, which are in rows and columns. By placing the pixels close together, it is hoped that the observer will perceive a continuous image. Obviously the finer the pixel spacing, the greater the resolution of the picture will

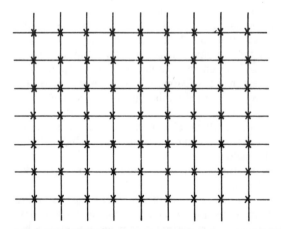

Figure 1.6(a) A picture can be stored digitally by representing the brightness at each of the above points by a binary number. For a colour picture each point becomes a vector and has to describe the brightness, hue and saturation of that part of the picture. Samples are usually but not always formed into regular arrays of rows and columns, and it is most efficient if the horizontal and vertical spacings are the same.

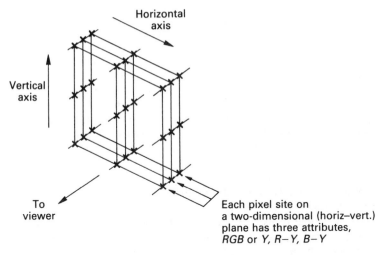

Horizontal
axis

Vertical
axis

To
viewer

Each pixel site on
a two-dimensional (horiz–vert.)
plane has three attributes,
RGB or *Y, R−Y, B−Y*

Figure 1.6(b) In the case of component video, each pixel site is described by three values and so the pixel becomes a vector quantity.

be, but the amount of data needed to store one picture will increase as the square of the resolution, and with it the costs.

If it is desired to convey a coloured image, then it will be seen from Figure 1.6(b) that each image consists of three superimposed layers of samples, one for each component. The pixel is no longer a single number representing a scalar brightness value, but a vector which describes in some way the brightness, hue and saturation of that point in the picture. In *RGB*, the pixels contain three unipolar (positive only) numbers representing the proportion of each of the three primary colours at that point in the picture. In colour difference working, each pixel again contains three numbers, but one of these is a unipolar number representing the luminance and the other two are bipolar (positive or negative) numbers representing the colour difference values.

In order to produce moving pictures, the current approach is simply to provide a mechanism where the value of every pixel can be updated periodically. This effectively results in a three-dimensional array, where two of the axes are spatial, and the third is temporal.

Since it is possible to represent any analog waveform digitally, composite video can also be digitized. Owing to the critical nature of chroma phase, the sampling rate will generally be locked to subcarrier so that it can be readily created from burst. It should be stressed that digital composite is simply another way of conveying composite video, and all of the attributes of composite, such as colour framing sequences, are still present in the resulting data.

At the ADC (analog-to-digital converter), every effort is made to rid the sampling clock of jitter, or time instability, so every sample is taken at an exactly even time step. Clearly if there is any subsequent timebase error, the instants at which samples arrive will be changed and the effect can be detected. If samples arrive at some destination with an irregular timebase, the effect can be eliminated by storing the samples temporarily in a memory and reading them out using a stable, locally generated clock. This process is called timebase correction and all properly engineered digital video systems must use it. Clearly timebase error is not reduced; it is totally eliminated. As a result there is little point measuring the timebase stability of a digital recorder.

Those who are not familiar with digital principles often worry that sampling takes away something from a signal because it is not taking notice of what happened between the samples. This would be true in a system having infinite bandwidth, but no analog signal can have infinite bandwidth. All analog signal sources from cameras, VTRs and so on have a resolution or frequency response limit, as indeed do devices such as CRTs and human vision. When a signal has finite bandwidth, the rate at which it can change is limited, and the way in which it changes becomes predictable. When a waveform can only change between samples in one way, it is then only necessary to convey the samples and the original waveform can be reconstructed from them. A more detailed treatment of the principle will be given in Chapter 2.

As stated, each sample is also discrete, or represented in a stepwise manner. The length of the sample, which will be proportional to the voltage of the video signal, is represented by a whole number. This process is known as quantizing and results in an approximation, but the size of the error can be controlled until it is negligible. If, for example, we were to measure the height of humans to the nearest metre, virtually all adults would register two metres high and obvious difficulties would result. These are generally overcome by measuring height to the nearest centimetre. Clearly there is no advantage in going further and expressing our height in a whole number of millimetres or even micrometres. The point is that an appropriate resolution can also be found for video signals, and a higher figure is not beneficial. The link between video quality and sample resolution is explored in Chapter 2. The advantage of using whole numbers is that they are not prone to drift. If a whole number can be carried from one place to another without numerical error, it has not changed at all. By describing video waveforms numerically, the original information has been expressed in a way which is better able to resist unwanted changes.

Essentially, digital video carries the original waveform numerically. The number of the sample is an analog of time, which itself is an analog of position across the screen, and the magnitude of the sample is (in the case of luminance) an analog of the brightness at the appropriate point in the image. In fact the succession of samples in a digital system is actually *an analog* of the original waveform. This sounds like a contradiction and as a result some authorities prefer the term 'numerical video' to 'digital video' and in fact the French word is *numérique*.

As both axes of the digitally represented waveform are discrete, the waveform can be accurately restored from numbers as if it were being drawn on graph paper. If we require greater accuracy, we simply choose paper with smaller squares. Clearly more numbers are then required and each one could change over a larger range.

In simple terms, the video waveform is conveyed in a digital recorder as if the voltage had been measured at regular intervals with a digital meter and the readings had been written down on a roll of paper. The rate at which the measurements were taken and the accuracy of the meter are the only factors which determine the quality, because once a parameter is expressed as a discrete number, a series of such numbers can be conveyed unchanged. Clearly in this example the handwriting used and the grade of paper have no effect on the information. The quality is determined only by the accuracy of conversion and is independent of the quality of the signal path.

1.4 Why binary?

Humans insist on using numbers expressed to the base of ten, having evolved with that number of digits. Other number bases exist; the most minimal system is binary, which has only two digits, 0 and 1. BInary digiTS are universally contracted to bits. These are readily conveyed in switching circuits by an 'on' state and an 'off' state. With only two states, there is little chance of error. Had humans evolved with only one finger, we might have adopted binary more commonly.

In decimal systems, the digits in a number (counting from the right, or least significant end) represent ones, tens, hundreds, thousands, etc. Figure 1.7 shows that in binary, the bits represent one, two, four, eight, sixteen, etc. A multidigit binary number is commonly called a word, and the number of bits in the word is called the wordlength. The right-hand bit is called the least significant bit (LSB) whereas the bit on the left-hand end of the word is called the most significant bit (MSB). Clearly more digits are required in binary than in decimal, but they are more easily handled. A word of eight bits is called a byte. The capacity of memories and storage media is measured in bytes, but to avoid large numbers, kilobytes, megabytes and gigabytes are often used. As memory addresses are themselves binary numbers, the wordlength limits the address range. The range is found by raising two to the power of the wordlength. Thus a 4 bit word has sixteen combinations, and could address a memory having sixteen locations. A 10 bit word has 1024 combinations, which is close to one thousand. In digital terminology, 1K = 1024, so a kilobyte of memory contains 1024 bytes. A megabyte (1MB) contains 1024 kilobytes and a gigabyte contains 1024 megabytes.

In a digital video system, the whole number representing the length of the sample is expressed in binary. The signals sent have two states, and change at

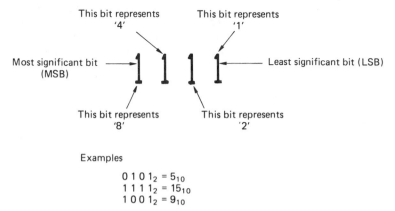

Examples

$$0\ 1\ 0\ 1_2 = 5_{10}$$
$$1\ 1\ 1\ 1_2 = 15_{10}$$
$$1\ 0\ 0\ 1_2 = 9_{10}$$

Figure 1.7 In a binary number, the digits represent increasing powers of two from the LSB. Also defined here are MSB and wordlength. When the wordlength is 8 bits, the word is a byte. Binary numbers are used as memory addresses, and the range is defined by the address wordlength. Some examples are shown here.

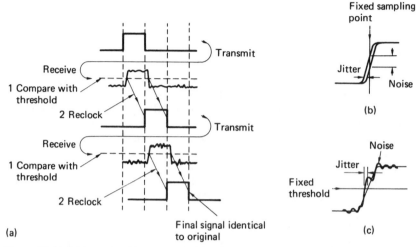

Figure 1.8 (a) A binary signal is compared with a threshold and relocked on receipt; thus the meaning will be unchanged. (b) Jitter on a signal can appear as noise with respect to fixed timing. (c) Noise on a signal can appear as jitter when compared with a fixed threshold.

predetermined times according to some stable clock. Figure 1.8 shows the consequences of this form of transmission. If the binary signal is degraded by noise, this will be rejected by the receiver, which judges the signal solely by whether it is above or below the half-way threshold, a process known as slicing. The signal will be carried in a channel with finite bandwidth, and this limits the slew rate of the signal; an ideally upright edge is made to slope. Noise added to a sloping signal can change the time at which the slicer judges that the level passed through the threshold. This effect is also eliminated when the output of the slicer is reclocked. However many stages the binary signal passes through, it still comes out the same, only later.

Video samples which are represented by whole numbers can be reliably carried from one place to another by such a scheme, and if the number is correctly received, there has been no loss of information *en route*.

There are two ways in which binary signals can be used to carry samples and these are shown in Figure 1.9. When each digit of the binary number is carried

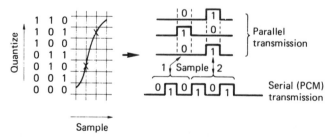

Figure 1.9 When a signal is carried in numerical form, either parallel or serial, the mechanisms of Fig. 1.8 ensure that the only degradation is in the conversion process.

on a separate wire this is called parallel transmission. The state of the wires changes at the sampling rate. Using multiple wires is cumbersome, particularly where a long wordlength is in use, and a single wire can be used where successive digits from each sample are sent serially. This is the definition of pulse code modulation. Clearly the clock frequency must now be higher than the sampling rate. Whilst the transmission of video by such a scheme is advantageous in that noise and timebase error have been eliminated, there is a penalty that a high-quality colour difference signal requires between two and three hundred million bits per second. Clearly digital video could only become commonplace when such a data rate could be handled economically. Further applications become possible when means to reduce the data rate become economic.

1.5 Why digital?

There are two main answers to this question, and it is not possible to say which is the most important, as it will depend on one's standpoint.

(1) The quality of reproduction of a well engineered digital video system is independent of the medium and depends only on the quality of the conversion processes.

(2) The conversion of video to the digital domain allows tremendous opportunities which were denied to analog signals.

Someone who is only interested in picture quality will judge the former the most relevant. If good-quality converters can be obtained, all of the shortcomings of analog recording or transmission can be eliminated to great advantage. One's greatest effort is expended in the design of converters, whereas those parts of the system which handle data need only be workmanlike. Timebase error, tape noise, print-through, dropouts, and moiré are all history. When a digital recording is copied, the same numbers appear on the copy: it is not a dub, it is a clone. If the copy is indistinguishable from the original, there has been no generation loss. Digital recordings can be copied indefinitely without loss of quality.

In the real world everything has a cost, and one of the greatest strengths of digital technology is low cost. If copying causes no quality loss, recorders do not need to be far better than necessary in order to withstand generation loss. They need only be adequate on the first generation whose quality is then maintained. There is no need for the great size and extravagant tape consumption of professional analog recorders. When the information to be recorded consists of discrete numbers, they can be packed densely on the medium without quality loss. Should some bits be in error because of noise or dropout, error correction can restore the original value. Digital recordings take up less space than analog recordings for the same or better quality. Tape costs are far less and storage costs are reduced.

Digital circuitry costs less to manufacture. Switching circuitry which handles binary can be integrated more densely than analog circuitry. More functionality can be put in the same chip. Analog circuits are built from a host of different component types which have a variety of shapes and sizes and are costly to assemble and adjust. Digital circuitry uses standardized component outlines and is easier to assemble on automated equipment. Little if any adjustment is needed.

Once video is in the digital domain, it becomes data, and as such is indistinguishable from any other type of data. Systems and techniques developed in other industries for other purposes can be used for video. Computer equipment is available at low cost because the volume of production is far greater than that of professional video equipment. Disk drives and memories developed for computers can be put to use in video products. A word processor adapted to handle video samples becomes a workstation. There seems to be little point in waiting for a tape to wind when a disk head can access data in milliseconds.

Communications networks developed to handle data can happily carry digital video and accompanying audio over indefinite distances without quality loss. Digital TV broadcasting makes use of these techniques to eliminate the interference, fading and multipath reception problems of analog broadcasting. At the same time, more efficient use is made of available bandwidth.

Digital equipment can have self-diagnosis programs built in. The machine points out its own failures. The days of chasing a signal with an oscilloscope are over. Even if a faulty component in a digital circuit could be located with such a primitive tool, it is well-nigh impossible to replace a chip having 60 pins soldered through a six-layer circuit board. The cost of finding the fault may be more than the board is worth. Routine, mind-numbing adjustment of analog circuits to counteract drift is no longer needed. The cost of maintenance falls. A small operation may not need maintenance staff at all; a service contract is sufficient. A larger organization will still need maintenance staff, but they will be fewer in number and their skills will be oriented more to systems than devices.

As a result of the above, the cost of ownership of digital equipment is less than that of analog. Debates about quality are academic; analog equipment can no longer compete economically, and it will eventually dwindle away as surely as the Compact Disc has now replaced vinyl.

1.6 Some digital video processes outlined

Whilst digital video is a large subject, it is not necessarily a difficult one. Every process can be broken down into smaller steps, each of which is relatively easy to follow. The main difficulty with study is to appreciate where the small steps fit in the overall picture. Subsequent chapters of this book will describe the key processes found in digital technology in more detail, whereas this chapter illustrates why these processes are necessary and shows how they are combined in various ways in real equipment. Once the general structure of digital devices is appreciated, the following chapters can be put in perspective.

Figure 1.10(a) shows a minimal digital video system. This is no more than a point-to-point link which conveys analog video from one place to another. It consists of a pair of converters and hardware to serialize and deserialize the samples. There is a need for standardization in serial transmission so that various devices can be connected together. These standards for digital interfaces are described in Chapter 5.

Analog video entering the system is converted in the analog-to-digital converter (ADC) to samples which are expressed as binary numbers. A typical sample would have a wordlength of 8 bits. The sample is connected in parallel into an output register which controls the cable drivers. The cable also carries the sampling-rate clock. The data are sent to the other end of the line where a slicer

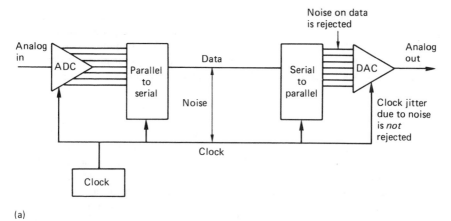

(a)

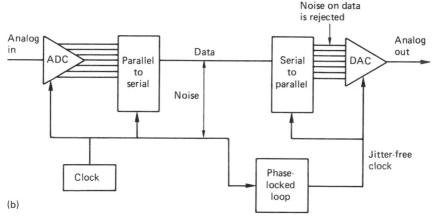

(b)

Figure 1.10 In (a) two converters are joined by a serial link. Although simple, this system is deficient because it has no means to prevent noise on the clock lines causing jitter at the receiver. In (b) a phase-locked loop is incorporated, which filters jitter from the clock.

rejects noise picked up on each signal. Sliced data are then loaded into a receiving register by the clock, and sent to the digital-to-analog converter (DAC), which converts the sample back to an analog voltage.

Following a casual study one might conclude that if the converters were of transparent quality, the system must be ideal. Unfortunately this is incorrect. As Figure 1.8 showed, noise can change the timing of a sliced signal. Whilst this system rejects noise which threatens to change the numerical value of the samples, it is powerless to prevent noise from causing jitter in the receipt of the word clock. Noise on the word clock means that samples are not converted with a regular timebase and the impairment caused can be noticeable. Stated another way, analog characteristics of the interconnect are not prevented from affecting the reproduced waveform and so the system is not truly digital.

The jitter problem is overcome in Figure 1.10(b) by the inclusion of a phase-locked loop which is an oscillator which synchronizes itself to the *average* frequency of the clock but which filters out the instantaneous jitter. The operation of a phase-locked loop is analogous to the function of the flywheel on a piston engine. The samples are then fed to the converter with a regular spacing and the impairment is no longer audible. Chapter 2 shows why the effect occurs and deduces the clock accuracy needed for accurate conversion.

1.7 The frame store

The system of Figure 1.10 is extended in Figure 1.11 by the addition of some random access memory (RAM). What the device does is determined by the way in which the RAM address is controlled. If the RAM address increases by 1 every time a sample from the ADC is stored in the RAM, a recording can be made for a short period until the RAM is full. The recording can be played back by repeating the address sequence at the same clock rate but reading the memory into the DAC. The result is generally called a frame store.[2] If the memory capacity is increased, the device can be used for recording. At a rate of 200 million bits per second, each frame needs a megabyte of memory and so the RAM recorder will be restricted to a fairly short playing time.

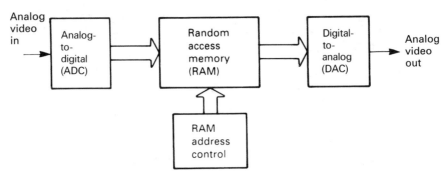

Figure 1.11 In the frame store, the recording medium is a random access memory (RAM). Recording time available is short compared with other media, but access to the recording is immediate and flexible as it is controlled by addressing the RAM.

Using data reduction, the playing time of a RAM-based recorder can be extended. For predetermined images such as test patterns and station IDs, read only memory (ROM) can be used instead as it is non-volatile.

1.8 The frame synchronizer

If the RAM is used in a different way, it can be written and read at the same time. The device then becomes a synchronizer which allows video interchange between two systems which are not genlocked. Controlling the relationship between the addresses makes the RAM into a variable delay. The addresses are generated by counters which overflow to zero after they have reached a maximum count at the end of a frame. As a result the memory space appears to be circular as shown in Figure 1.12. The read and write addresses chase one another around the circle. If the read address follows close behind the write address, the delay is short. If it just stays ahead of the write address, the

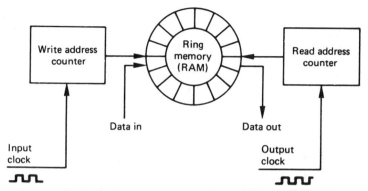

Figure 1.12 If the memory address is arranged to come from a counter which overflows, the memory can be made to appear circular. The write address then rotates endlessly, overwriting previous data once per revolution. The read address can follow the write address by a variable distance (not exceeding one revolution) and so a variable delay takes place between reading and writing.

maximum delay is reached. If the input and output have an identical frame rate, the address relationship will be constant, but if there is a drift then the address relationship will change slowly. Eventually the addresses will coincide and then cross. Properly handled, this results in a frame being omitted or repeated.

A synchronizer with less than a frame of RAM can be used to remove static timing errors due, for example, to propagation delays in large systems. The finite RAM capacity gives a finite range of timing error which can be accommodated. This is known as the window.

1.9 The DVE

The modern digital video effects (DVE) unit has a large repertoire of manipulations. These can be divided into various categories. Certain effects such

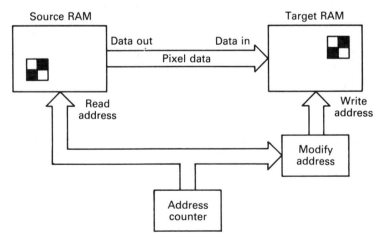

Figure 1.13 A DVE manipulates images by transferring from one memory to another but using different read and write addresses.

as posterization or mosaicing do not move the picture and are quite simple to perform. More complex effects manipulate the size and position of the picture. Figure 1.13 shows that these are achieved by transferring pixel data from one frame store to another. The read and write addresses are different so that pixel values move across the screen. There are a number of difficulties to overcome, such as the change in pixel spacing when the size is changed, and the solution to this is a form of sampling-rate conversion. Details of the process are considered in Chapter 3.

1.10 Time compression

When samples are converted, the ADC must run at a constant clock rate and it outputs an unbroken stream of samples during the active line. Following a break during blanking, the sample stream resumes. Time compression allows the sample stream to be broken into blocks for convenient handling.

Figure 1.14 shows an ADC feeding a pair of RAMs. When one is being written by the ADC, the other can be read, and vice versa. As soon as the first RAM is full, the ADC output switches to the input of the other RAM so that there is no loss of samples. The first RAM can then be read at a higher clock rate than the sampling rate. As a result the RAM is read in less time than it took to write it, and the output from the system then pauses until the second RAM is full. The samples are now time compressed. Instead of being an unbroken stream which is difficult to handle, the samples are now arranged in blocks with convenient pauses in between them. In these pauses numerous processes can take place. A rotary-head recorder might spread the data from a field over several tape tracks; a hard disk might move to another track. In all types of recording, the time compression of the samples allows time for synchronizing patterns, subcode and error-correction words to be recorded.

In digital VTRs, the video data are time compressed so that part of the track is left for audio data. Figure 1.15 shows that heavy time compression of the audio data raises the data rate up to that of the video data so that the same tracks, heads and much common circuitry can be used to record both.

Subsequently, any time compression can be reversed by time expansion. Samples are written into a RAM at the incoming clock rate, but read out at the standard sampling rate. Unless there is a design fault, time compression is totally undetectable. In a recorder, the time-expansion stage can be combined with the timebase-correction stage so that speed variations in the medium can be eliminated at the same time. The use of time compression is universal in digital recording. In general the *instantaneous* data rate at the medium is not the same as the rate at the converters, although clearly the *average* rate must be the same.

Another application of time compression is to allow several channels of audio to be carried along with video on a single cable. This technique is used in the serial digital interface (SDI) which is explained in Chapter 5.

1.11 Synchronization

The issue of signal timing has always been critical in analog video, particularly in composite working, but the adoption of digital routing relaxes the requirements considerably. Analog vision mixers need to be fed by equal-length

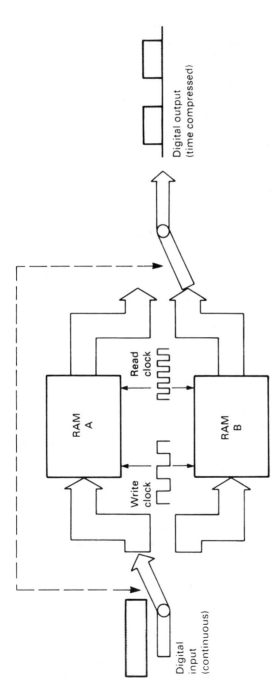

Figure 1.14 In time compression, the unbroken real-time stream of samples from an ADC is broken up into discrete blocks. This is accomplished by the configuration shown here. Samples are written into one RAM at the sampling rate by the write clock. When the first RAM is full, the switches change over, and writing continues into the second RAM whilst the first is read using a higher-frequency clock. The RAM is read faster than it was written and so all of the data will be output before the other RAM is full. This opens spaces in the data flow which are used as described in the text.

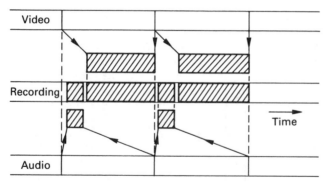

Figure 1.15 Time compression is used to shorten the length of track needed by the video. Heavily time-compressed audio samples can then be recorded on the same track using common circuitry.

cables from the router to prevent propagation delay variation. In the digital domain this is no longer an issue as delay is easily obtained and each input of a digital vision mixer can have its own local timebase corrector. Provided signals are received having timing within the window of the inputs, all inputs are retimed to the same phase within the mixer. Chapter 5 deals with synchronizing large systems.

1.12 Error correction and concealment

As anyone familiar with analog recording will know, magnetic tape is an imperfect medium. It suffers from noise and dropouts, which in analog recording are visible. In a digital recording of binary data, a bit is either correct or wrong, with no intermediate stage. Small amounts of noise are rejected, but inevitably, infrequent noise impulses cause some individual bits to be in error. Dropouts cause a larger number of bits in one place to be in error. An error of this kind is called a burst error. Whatever the medium and whatever the nature of the mechanism responsible, data either are recovered correctly, or suffer some combination of bit errors and burst errors. In optical disks, random errors can be caused by imperfections in the moulding process, whereas burst errors are due to contamination or scratching of the disk surface.

The visibility of a bit error depends upon which bit of the sample is involved. If the LSB of one sample was in error in a detailed, contrast picture, the effect would be totally masked and no-one could detect it. Conversely, if the MSB of one sample was in error in a flat field, no-one could fail to notice the resulting spot. Clearly a means is needed to render errors from the medium invisible. This is the purpose of error correction.

In binary, a bit has only two states. If it is wrong, it is only necessary to reverse the state and it must be right. Thus the correction process is trivial and perfect. The main difficulty is in identifying the bits which are in error. This is done by coding the data by adding redundant bits. Adding redundancy is not confined to digital technology: airliners have several engines and cars have twin braking systems. Clearly the more failures which have to be handled, the more redundancy is needed. If a four-engined airliner is designed to fly normally with one engine failed, three of the engines have enough power to reach cruise speed, and the fourth one is redundant. The amount of redundancy is equal to the amount of failure which can be handled. In the case of the failure of two engines, the plane can still fly, but it must slow down; this is graceful degradation. Clearly the chances of a two-engine failure on the same flight are remote.

In digital recording, the amount of error which can be corrected is proportional to the amount of redundancy, and it will be shown in Chapter 4 that within this limit, the samples are returned to exactly their original value. Consequently *corrected* samples are undetectable. If the amount of error exceeds the amount of redundancy, correction is not possible, and, in order to allow graceful degradation, concealment will be used. Concealment is a process where the value of a missing sample is estimated from those nearby. The estimated sample value is not necessarily exactly the same as the original, and so under some circumstances concealment can be audible, especially if it is frequent. However, in a well-designed system, concealments occur with negligible frequency unless there is an actual fault or problem.

Concealment is made possible by rearranging the sample sequence prior to recording. This is shown in Figure 1.16 where odd-numbered samples are separated from even-numbered samples prior to recording. The odd and even sets of samples may be distributed over two tape tracks, so that an uncorrectable burst error only affects one set. On replay, the samples are recombined into their natural sequence, and the error is now split up so that it results in every other sample being lost in a two-dimensional structure. The picture is now described half as often, but can still be reproduced with some loss of accuracy. This is better than not being reproduced at all even if it is not perfect. Almost all digital

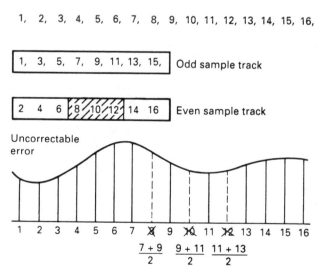

Figure 1.16 In cases where the error correction is inadequate, concealment can be used provided that the samples have been ordered appropriately in the recording. Odd and even samples are recorded in different places as shown here. As a result an uncorrectable error causes incorrect samples to occur singly, between correct samples. In the example shown, sample 8 is incorrect, but samples 7 and 9 are unaffected and an approximation to the value of sample 8 can be had by taking the average value of the two. This interpolated value is substituted for the incorrect value.

recorders use such an odd/even distribution for concealment. Clearly if any errors are fully correctable, the distribution is a waste of time; it is only needed if correction is not possible.

The presence of an error-correction system means that the video (and audio) quality is independent of the tape/head quality within limits. There is no point in trying to assess the health of a machine by watching a monitor or listening to the audio, as this will not reveal whether the error rate is normal or within a whisker of failure. The only useful procedure is to monitor the frequency with which errors are being corrected, and to compare it with normal figures. Professional DVTRs have an error rate display for this purpose and in addition most allow the error-correction system to be disabled for testing.

1.13 Product codes

In high-density recorders, more data are lost in a given-sized dropout. Adding redundancy equal to the size of a dropout to every code is inefficient. Figure 1.17(a) shows that the efficiency of the system can be raised using interleaving. Following distribution, sequential samples from the ADC are assembled into codes, but these are not recorded in their natural sequence. A number of sequential codes are assembled along rows in a memory. When the memory is full, it is copied to the medium by reading down columns. On replay, the samples need to be de-interleaved to return them to their natural sequence. This is done by writing samples from tape into a memory in columns, and when it is full, the memory is read in rows. Samples read from the memory are now in their original sequence so there is no effect on the recording. However, if a burst error occurs on the medium as is shown shaded in the diagram, it will damage sequential samples in a vertical direction in the de-interleave memory. When the memory is read, a single large error is broken down into a number of small errors whose size is exactly equal to the correcting power of the codes and the correction is performed with maximum efficiency.

An extension of the process of interleave is where the memory array has not only rows made into codewords, but also columns made into codewords by the addition of vertical redundancy. This is known as a product code. Figure 1.17(b) shows that in a product code the redundancy calculated first and checked last is called the outer code, and the redundancy calculated second and checked first is called the inner code. The inner code is formed along tracks on the medium. Random errors due to noise are corrected by the inner code and do not impair the burst-correcting power of the outer code. Burst errors are declared uncorrectable by the inner code which flags the bad samples on the way into the de-interleave memory. The outer code reads the error flags in order to locate the erroneous data. As it does not have to compute the error locations, the outer code can correct more errors.

The interleave, de-interleave, time-compression and timebase-correction processes cause delay and this is evident in the timing of the confidence replay output of DVTRs.

1.14 Shuffling

When a product-code-based recording suffers an uncorrectable error the result is a rectangular block of failed sample values which require concealment. Such a

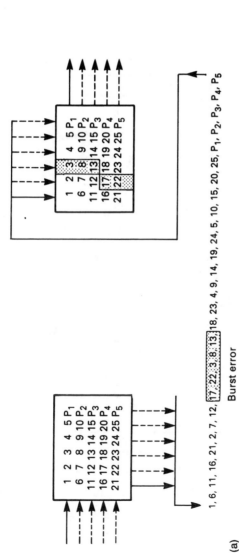

1, 6, 11, 16, 21, 2, 7, 12, 17, 22, 3, 8, 13, 18, 23, 4, 9, 14, 19, 24, 5, 10, 15, 20, 25, P₁, P₂, P₃, P₄, P₅

Burst error

(a)

Figure 1.17(a) Interleaving is essential to make error-correction schemes more efficient. Samples written sequentially in rows into a memory have redundancy P added to each row. The memory is then read in columns and the data sent to the recording medium. On replay the non-sequential samples from the medium are de-interleaved to return them to their normal sequence. This breaks up the burst error (shaded) into one error symbol per row in the memory, which can be corrected by the redundancy P.

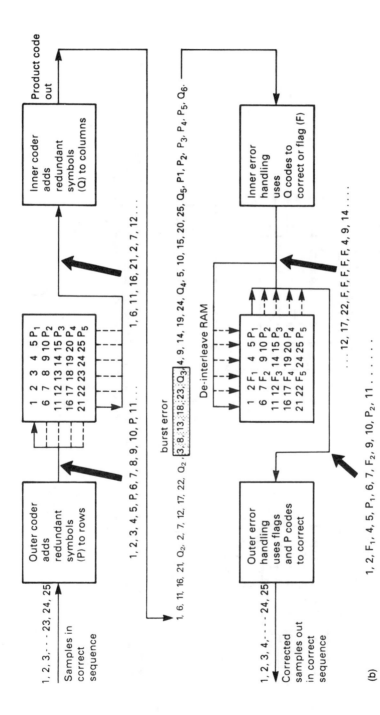

Figure 1.17(b) In addition to the redundancy P on rows, inner redundancy Q is also generated on columns. On replay, the Q code checker will pass on flags F if it finds an error too large to handle itself. The flags pass through the de-interleave process and are used by the outer error correction to identify which symbol in the row needs correcting with P redundancy. The concept of crossing two codes in this way is called a product code.

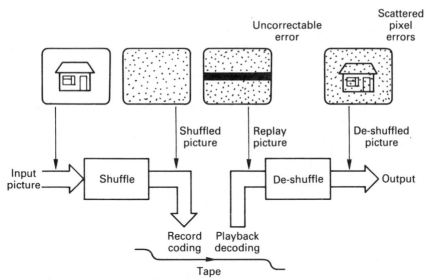

Figure 1.18 The shuffle before recording and the corresponding de-shuffle after playback cancel out as far as the picture is concerned. However, a block of errors due to dropout only experiences the de-shuffle, which spreads the error randomly over the screen. The pixel errors are then easier to correct.

regular structure would be visible even after concealment, and an additional process is necessary to reduce the visibility. Figure 1.18 shows that a shuffle process is performed prior to product coding in which the pixels are moved around the picture in a pseudo-random fashion. The reverse process is used on replay, and the overall effect is nullified. However, if an uncorrectable error occurs, this will only pass through the de-shuffle and so the regular structure of the failed data blocks will be randomized. The errors are spread across the picture as individual failed pixels in an irregular structure.

1.15 Channel coding

In most recorders used for storing digital information, the medium carries a track which reproduces a single waveform. Clearly data words representing video samples contain many bits and so they have to be recorded serially, a bit at a time. Some media, such as disks, only have one active track, so it must be totally self-contained. DVTRs usually have two or four tracks read or written on simultaneously. At high recording densities, physical tolerances cause phase shifts, or timing errors, between tracks and so it is not possible to read them in parallel. Each track must still be self-contained until the replayed signal has been timebase corrected.

Recording data serially is not as simple as connecting the serial output of a shift register to the head. In digital video, samples may contain strings of identical bits. If a shift register is loaded with such a sample and shifted out serially, the output stays at a constant level for the period of the identical bits, and nothing is recorded on the track. On replay there is nothing to indicate how many bits were present, or even how fast to move the medium. Clearly serialized raw data cannot be recorded directly; they must be modulated into a waveform which contains an embedded clock irrespective of the values of the bits in the samples. On replay a circuit called a data separator can lock to the embedded clock and use it to separate strings of identical bits.

The process of modulating serial data to make them self-clocking is called channel coding. Channel coding also shapes the spectrum of the serialized waveform to make it more efficient. With a good channel code, more data can be stored on a given medium. Spectrum shaping is used in optical disks to prevent the data from interfering with the focus and tracking servos, and in certain DVTRs to allow re-recording without erase heads.

Channel coding is needed for serial digital transmission in cables and is also needed to broadcast digital television signals where shaping of the spectrum is an obvious requirement to avoid interference with other services. NICAM TV sound and digital audio broadcasting (DAB) rely on it.

The techniques of channel coding are introduced in Chapter 4.

1.16 Standards conversion

In addition to changing the number of lines in the picture, standards conversion requires new fields to be created on the time axis in between input fields, as shown in Figure 1.19. Both of these processes are a form of sampling-rate conversion and are performed with logic similar to that contained in a DVE. The field-rate conversion requires interpolation along the time axis using several fields, typically four, as input. Objects which are moving appear in different places in each field and with a conventional standards converter the result is judder or double images in the output. Until recently these motion artefacts had to be accepted, but now solutions are available at the cost of some complexity.

Motion estimation is a technique for measuring how objects in a television picture move from one image to another. The movement is expressed as a series of motion vectors which have both direction and magnitude. Motion compensation is a technique which uses the motion vectors to enhance some process. One of the first applications is in motion-compensated standards conversion which identifies moving objects and measures their displacement from field to field. An opposing, or compensating, displacement can then be used in each input field to align each object with its location in the output field. The result is freedom from judder and better resolution of moving objects.

1.17 Video data reduction

Digital video suffers from an extremely high data rate, particularly in high definition, and one approach to the problem is to reduce that rate without affecting the subjective quality of the picture. The human eye is not equally sensitive to all spatial frequencies, so some coding gain can be obtained by quantizing more coarsely the frequencies which are less visible. Video images typically contain a great deal of redundancy where flat areas contain the same pixel value repeated many times. Furthermore, in many cases there is little difference between one field and the next, and interfield data reduction can be achieved by sending only the differences. Whilst this may achieve considerable reduction, the result is difficult to edit because individual fields can no longer be

Time

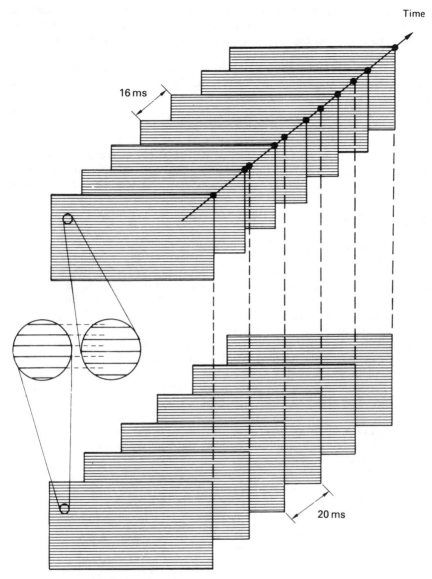

16 ms

20 ms

Figure 1.19 The basic problem of standards conversion. The line spacing is easy to accommodate. The field spacing is not.

identified in the data stream. Thus for production purposes, data reduction is restricted to exploiting the redundancy within each field individually. Production DVTRs such as Sony's Digital Betacam and the Ampex DCT use only very mild compression of 2:1. This allows simple algorithms to be used and also permits multiple generations without artefacts being visible. A similar approach is used in disk-based workstations such as Lightworks and AVID. Data reduction is discussed in Chapter 3.

Clearly a consumer DVTR needs only single-generation operation and has simple editing requirements. A much greater degree of compression can then be used, which also takes advantage of redundancy between fields. The same is true for broadcasting, where bandwidth is at a premium. A similar approach may be used in disk-based camcorders which are intended for ENG purposes.

There is now no doubt that the future of television broadcasting (and radio for that matter) lies in data-reduced digital technology.

Data reduction requires an encoder prior to the recording medium and a decoder after it. Unless the correct decoder is available, the recording cannot be played. Interchange of data-reduced material will be hampered unless coding standards are available. Data reduction and the corresponding decoding are complex processes and take time, adding to existing delays in signal paths. Concealment of uncorrectable errors is also more difficult on reduced data.

1.18 Disk-based recording

The magnetic disk drive was perfected by the computer industry to allow rapid random access to data, and so it makes an ideal medium for editing. As will be seen in Chapter 7, the heads do not touch the disk, but are supported on a thin air film which gives them a long life but which restricts the recording density. Thus disks cannot compete with tape for lengthy recordings, but for short-duration work such as commercials or animation they have no equal. The data rate of digital video is too high for a single disk head, and so a number of solutions have been explored. One obvious solution is to use data reduction, which cuts the data rate and extends the playing time. Another approach is to use special multiplatter disk drives in which each disk head has its own circuitry allowing parallel transfer. Yet another possibility is to operate a large array of conventional drives in parallel. The highest-capacity magnetic disks are not removable from the drive.

Development of the optical disk was stimulated by the availability of low-cost lasers. Optical disks are available in many different types, some of which can

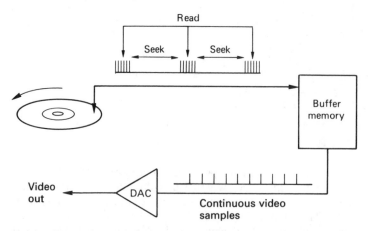

Figure 1.20 In a hard-disk recorder, a large-capacity memory is used as a buffer or timebase corrector between the converters and the disk. The memory allows the converters to run constantly despite the interruptions in disk transfer caused by the head moving between tracks.

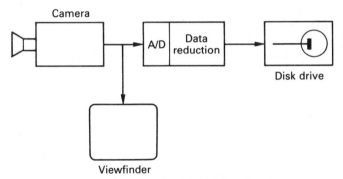

Figure 1.21 In a disk-based camcorder, PCM data rate from the camera is too high for direct recording on disk. Data reduction is used to cut the bit rate and extend playing time. If standard file structure is used, disks may be physically transferred to an edit system after recording.

only be recorded once, some of which are erasable. These will be contrasted in Chapter 7. Optical disks have in common the fact that access is generally slower than with magnetic drives and it is difficult to obtain high data rates, but most of them are removable and can act as interchange media.

The disk drive suffers from intermittent data transfer owing to the need to reposition the heads. Figure 1.20 shows that disk-based devices rely on a quantity of RAM acting as a buffer between the real-time video environment and the intermittent data environment.

Figure 1.21 shows the block diagram of a camcorder based on hard disks and data reduction. The recording time and picture quality will not compete with full-bandwidth tape-based devices, but following acquisition the disks can be used directly in an edit system, allowing a useful time saving in ENG applications.

1.19 Rotary-head digital recorders

The rotary-head recorder has the advantage that the spinning heads create a high head-to-tape speed offering a high bit rate recording without high tape speed. Whilst mechanically complex, the rotary-head transport has been raised to a high degree of refinement and offers the highest recording density and thus lowest cost per bit of all digital recorders.[3] Rotary-head transports are considered in Chapter 6.

Digital VTRs segment incoming fields into several tape tracks and invisibly reassemble them in memory on replay in order to keep the tracks reasonably short.

Figure 1.22 shows a representative block diagram of a DVTR. Following the converters, an optional data reduction unit may be found. There will be distribution of odd and even samples and a shuffle process for concealment purposes. An interleaved product code will be formed prior to the channel coding stage which produces the recorded waveform. On replay the data separator decodes the channel code and the inner and outer codes perform correction as in Section 1.11. Following the de-shuffle the data channels are recombined and any necessary concealment will take place. Any data expansion necessary will be performed prior to the output converters.

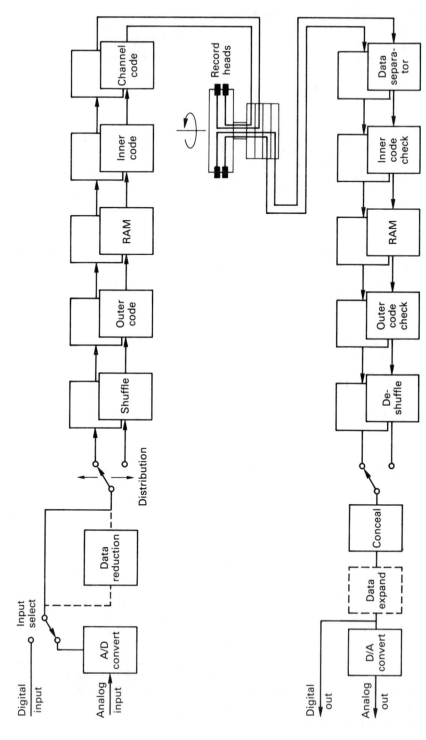

Figure 1.22 Block diagram of a DVTR. Note optional data reduction unit which may be used to allow a common transport to record a variety of formats.

1.20 Digital broadcasting

Although it has given good service for many years, analog television broadcasting is extremely inefficient because the transmitted waveform is directly compatible with the CRT display, and nothing is transmitted during the blanking periods whilst the beam retraces. Using time compression, data reduction and digital modulation techniques the same picture quality can be obtained in a fraction of the bandwidth of analog. Pressure on spectrum use from other users such as mobile telephones will only increase and this will ensure that future television and radio broadcasts will be digital.

If high-definition television ever reaches the home it will do so via digital data-reduced transmission as the bandwidth required will otherwise be hopelessly uneconomic. In addition to conserving spectrum, digital transmission is resistant to multipath reception and gives consistent picture quality throughout the service area.

References

1. DEVEREUX, V.G., Pulse code modulation of video signals: 8 bit coder and decoder. *BBC Res. Dept. Rept.*, **EL-42**, No.25 (1970)
2. PURSELL, S. and NEWBY, H., Digital frame store for television video. *SMPTE J.*, **82**, 402–403 (1973)
3. BALDWIN, J.L.E., Digital television recording – history and background. *SMPTE J.*, **95**, 1206–1214 (1986)

Chapter 2

Conversion

The power of digital techniques is in storage without degradation and in complex processing and manipulations. If video is to take advantage of digital techniques, then means must be found to express the colour and movement of a real-life scene as a series of numbers. Once in the digital domain, degradations can be controlled, and so the conversion processes to and from the digital domain become the main sources of degradation.

2.1 Introduction to conversion

There are a number of ways in which a video waveform can be digitally represented, but the most useful and therefore common is pulse code modulation or PCM which was introduced in Chapter 1. The input is a continuous-time, continuous-voltage video waveform, and this is converted into a discrete-time, discrete-voltage format by a combination of sampling and quantizing. As these two processes are orthogonal (a 64 dollar word for at right angles to one another) they are totally independent and can be performed in either order. Figure 2.1(a) shows an analog sampler preceding a quantizer, whereas (b) shows an asynchronous quantizer preceding a digital sampler. Ideally, both will give the same results; in practice each has different advantages and suffers from different deficiencies. The second approach is more common in video equipment.

The independence of sampling and quantizing allows each to be discussed quite separately in some detail, prior to combining the processes for a full understanding of conversion.

Whilst sampling an analog video waveform takes place in the time domain in an electrical ADC; this is because the analog waveform is the result of scanning an image. In reality the image has been spatially sampled in two dimensions (lines and pixels) and temporally sampled into fields along a third dimension. Sampling in a single dimension will be considered before moving on to more dimensions.

2.2 Sampling and aliasing

Sampling is no more than periodic measurement, and it will be shown here that there is no theoretical need for sampling to be detectable. However, practical television equipment often falls short of the ideal, particularly in the case of temporal sampling.

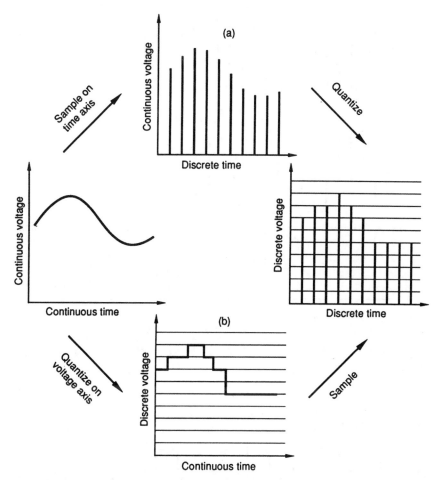

Figure 2.1 Since sampling and quantizing are orthogonal, the order in which they are performed is not important. In (a) sampling is performed first and the samples are quantized. This is common in audio converters. In (b) the analog input is quantized into an asynchronous binary code. Sampling takes place when this code is latched on sampling clock edges. This approach is universal in video converters.

Video sampling must be regular, because the process of timebase correction prior to conversion back to a conventional analog waveform assumes a regular original process as was shown in Chapter 1. The sampling process originates with a pulse train which is shown in Figure 2.2(a) to be of constant amplitude and period. The video waveform amplitude-modulates the pulse train in much the same way as the carrier is modulated in an AM radio transmitter. One must be careful to avoid over-modulating the pulse train as shown in (b) and this is achieved by applying a DC offset to the analog waveform so that blanking corresponds to a level part-way up the pulses as in (c).

In the same way that AM radio produces sidebands or images above and below the carrier, sampling also produces sidebands although the carrier is now a pulse train and has an infinite series of harmonics as shown in Figure 2.3(a). The

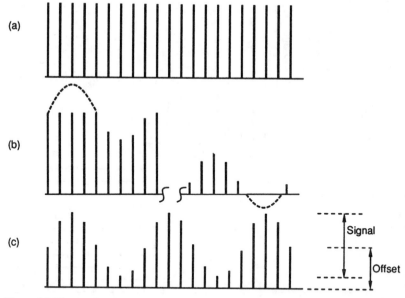

Figure 2.2 The sampling process requires a constant-amplitude pulse train as shown in (a). This is amplitude modulated by the waveform to be sampled. If the input waveform has excessive amplitude or incorrect level, the pulse train clips as shown in (b). For a bipolar waveform, the greatest signal level is possible when an offset of half the pulse amplitude is used to centre the waveform as shown in (c).

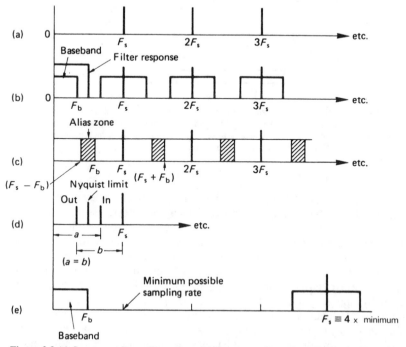

Figure 2.3 (a) Spectrum of sampling pulses. (b) Spectrum of samples. (c) Aliasing due to sideband overlap. (d) Beat-frequency production. (e) 4× oversampling.

sidebands repeat above and below each harmonic of the sampling rate as shown in (b).

The sampled signal can be returned to the continuous-time domain simply by passing it into a low-pass filter. This filter has a frequency response which prevents the images from passing, and only the baseband signal emerges, completely unchanged. If considered in the frequency domain, this filter can be called an anti-image filter; if considered in the time domain it can be called a reconstruction filter. It can also be considered as a spatial filter if a sampled still image is being returned to a continuous image. Such a filter will be two dimensional.

If an input is supplied having an excessive bandwidth for the sampling rate in use, the sidebands will overlap (Figure 2.3(c)) and the result is aliasing, where certain output frequencies are not the same as their input frequencies but instead become difference frequencies (Figure 2.3(d)). It will be seen from Figure 2.3 that aliasing does not occur when the input frequency is equal to or less than half the sampling rate, and this derives the most fundamental rule of sampling, which is that the sampling rate must be at least twice the highest input frequency. Sampling theory is usually attributed to Shannon[1,2] who applied it to information theory at around the same time as Kotelnikov in Russia. These applications were pre-dated by Whittaker. Despite that it is often referred to as Nyquist's theorem.

Whilst aliasing has been described above in the frequency domain, it can be described equally well in the time domain. In Figure 2.4(a) the sampling rate is obviously adequate to describe the waveform, but in (b) it is inadequate and aliasing has occurred.

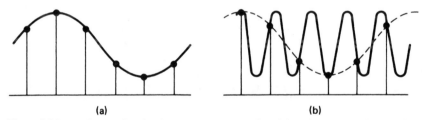

(a)   **(b)**

Figure 2.4 In (a), the sampling is adequate to reconstruct the original signal. In (b) the sampling rate is inadequate, and reconstruction produces the wrong waveform (dotted). Aliasing has taken place.

One often has no control over the spectrum of input signals and in practice it is necessary also to have a low-pass filter at the input to prevent aliasing. This anti-aliasing filter prevents frequencies of more than half the sampling rate from reaching the sampling stage. The requirement for an anti-aliasing filter extends to two-dimensional sampling devices such as CCD sensors, as will be seen in Section 2.6.

Whilst electrical or optical anti-aliasing filters are quite feasible, there is no corresponding device which can precede the image sampling at frame or field rate in film or TV cameras and as a result aliasing is commonly seen on television and in the cinema, owing to the relatively low frame rates used. With a frame rate of 24 Hz, a film camera will alias on any object changing at more than 12 Hz. Such objects include the spokes of stagecoach wheels; when the spoke-passing

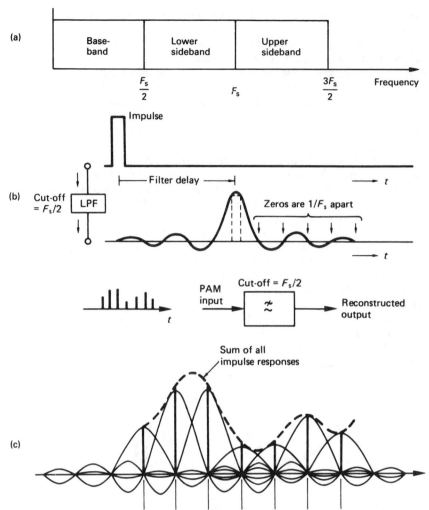

Figure 2.5 If ideal 'brick wall' filters are assumed, the efficient spectrum of (a) results. An ideal low-pass filter has an impulse response shown in (b). The impulse passes through zero at intervals equal to the sampling period. When convolved with a pulse train at the sampling rate, as shown in (c), the voltage at each sample instant is due to that sample alone as the impulses from all other samples pass through zero there.

frequency reaches 24 Hz the wheels appear to stop. Temporal aliasing in television is less visible than might be thought because of the way in which the eye perceives motion. This will be discussed further in Section 2.5.

If ideal low-pass anti-aliasing and anti-image filters are assumed, having a vertical cut-off slope at half the sampling rate, an ideal spectrum shown in Figure 2.5(a) is obtained. It can be seen from Figure 2.5(b) that the impulse response of a phase-linear ideal low-pass filter is a $\sin x/x$ waveform in the time domain. Such a waveform passes through zero volts periodically. If the cut-off frequency of the filter is one-half of the sampling rate, the impulse passes through zero *at the sites of all other samples.* It can be seen from Figure 2.5(c) that at the output of such a filter, the voltage at the centre of a sample is due to that sample alone, since the value of *all* other samples is zero at that instant. In other words the continuous-time output waveform must join up the tops of the input samples. In between the sample instants, the output of the filter is the sum of the contributions from many impulses, and the waveform smoothly joins the tops of the samples. If the time domain is being considered, the anti-image filter of the frequency domain can equally well be called the reconstruction filter. It is a consequence of the band-limiting of the original anti-aliasing filter that the filtered analog waveform could only travel between the sample points in one way. As the reconstruction filter has the same frequency response, the reconstructed output waveform must be identical to the original band-limited waveform prior to sampling.

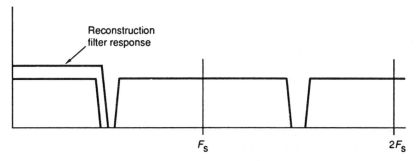

Figure 2.6 As filters with finite slope are needed in practical systems, the sampling rate is raised slightly beyond twice the highest frequency in the baseband.

The ideal filter with a vertical 'brick-wall' cut-off slope is difficult to implement. As the slope tends to the vertical, the delay caused by the filter goes to infinity. In practice, a filter with a finite slope has to be accepted as shown in Figure 2.6. The cut-off slope begins at the edge of the required band, and consequently the sampling rate has to be raised a little to drive aliasing products to an acceptably low level. There is no absolute factor by which the sampling rate must be raised; it depends upon the filters which are available and the level of aliasing products which are acceptable. The latter will depend upon the wordlength to which the signal will be quantized.

2.3 Aperture effect

The reconstruction process of Figure 2.5 only operates exactly as shown if the impulses are of negligible duration. In many processes this is not the case, and many real devices keep the analog signal constant for a substantial part of or even all the period. The result is a waveform which is more like a staircase than a pulse train. The case where the pulses have been extended in width to become equal to the sample period is known as a zero-order hold system and has a 100% aperture ratio.

Pulses of negligible width have a uniform spectrum which is flat within the baseband, while pulses of 100% aperture ratio have a $\sin x/x$ spectrum which is shown in Figure 2.7. The frequency response falls to a null at the sampling rate,

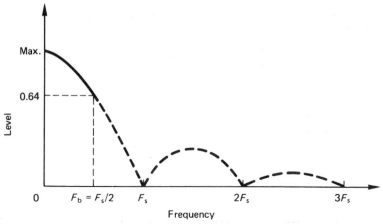

Figure 2.7 Frequency response with 100% aperture nulls at multiples of sampling rate. Area of interest is up to half sampling rate.

and as a result is about 4 dB down at the edge of the baseband. If the pulse width is stable, the reduction of high frequencies is constant and predictable, and an appropriate equalization circuit can render the overall response flat once more. An alternative is to use resampling which is shown in Figure 2.8. Resampling passes the zero-order hold waveform through a further synchronous sampling stage which consists of an analog switch which closes briefly in the centre of each sample period. The output of the switch will be pulses which are narrower than the original. If, for example, the aperture ratio is reduced to 50% of the sample period, the first frequency response null is now at twice the sampling rate, and the loss at the edge of the audio band is reduced. As the figure shows, the frequency response becomes flatter as the aperture ratio falls. The process should not be carried too far, as with very small aperture ratios there is little energy in the pulses and noise can be a problem. A practical limit is around 12.5% where the frequency response is virtually ideal.

The aperture effect will show up in many aspects of television. Lenses have finite MTF (modulation transfer function), such that a very small object becomes spread in the image. The image sensor will also have a finite aperture function. In tube cameras, the beam will have a finite radius, and will not necessarily have a uniform energy distribution across its diameter. In CCD cameras, the sensor is split into elements which may almost touch in some cases. The element integrates light falling on its surface, and so will have a rectangular aperture. In both cases there will be a roll-off of higher spatial frequencies.

As noted, in conventional tube cameras and CRTs the horizontal dimension is continuous, whereas the vertical dimension is sampled. The aperture effect means that the vertical resolution in real systems will be less than sampling theory permits, and to obtain equal horizontal and vertical resolutions a greater number of lines is necessary. The magnitude of the increase is described by the so-called Kell factor,[3] although the term factor is a misnomer since it can have a range of values depending on the apertures in use and the methods used to measure resolution. In digital video, sampling takes place in horizontal and vertical

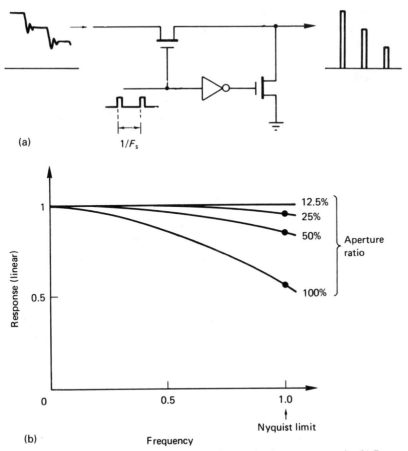

(a)

$1/F_s$

(b)   Frequency

Nyquist limit

Response (linear)

Figure 2.8 (a) Resampling circuit eliminates transients and reduces aperture ratio. (b) Response of various aperture ratios.

dimensions, and the Kell parameter becomes unnecessary. The outputs of digital systems will, however, be displayed on raster scan CRTs, and the Kell parameter of the display will then be effectively in series with the other system constraints.

The temporal aperture effect varies according to the equipment used. Tube cameras have a long integration time and thus a wide temporal aperture. Whilst this reduces temporal aliasing, it causes smear on moving objects. CCD cameras do not suffer from lag and as a result their temporal response is better. Some CCD cameras deliberately have a short temporal aperture as the time axis is resampled by a mechanically driven revolving shutter. The intention is to reduce smear (hence the popularity of such devices for sporting events) but there will be more aliasing on certain subjects. The effect of this will be considered in Section 2.5.

The eye has a temporal aperture effect which is known as persistence of vision, and the phosphors of CRTs continue to emit light after the electron beam has passed. These produce further temporal aperture effects in series with those in the camera. Current liquid crystal displays do not generate light, but act as a modulator to a separate light source. Their temporal response is rather slow, but there is a possibility to resample by pulsing the light source.

2.4 Two-dimensional sampling

Analog video samples in the time domain and vertically, whereas a two-dimensional still image such as a photograph must be sampled horizontally and vertically. In both cases a two-dimensional spectrum will result, one vertical/temporal and one vertical/horizontal.

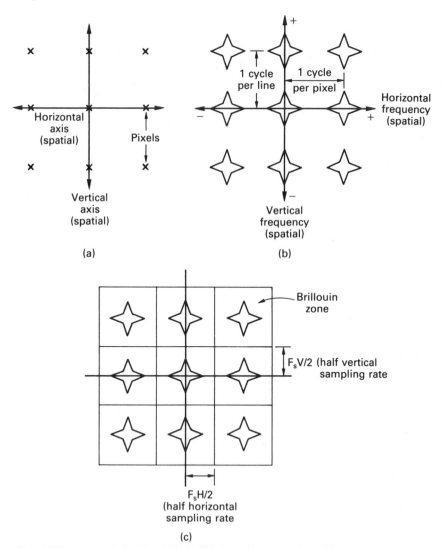

Figure 2.9 Image sampling spectra. The rectangular array of (a) has a spectrum shown in (b) having a rectangular repeating structure. Filtering to return to the baseband requires a two-dimensional filter whose response lies within the Brillouin zone shown in (c).

Figure 2.9(a) shows a square matrix of sampling sites which has an identical spatial sampling frequency both vertically and horizontally. The corresponding spectrum is shown in Figure 2.9(b). The baseband spectrum is in the centre of the diagram, and the repeating sampling sideband spectrum extends vertically and horizontally. The star-shaped spectrum results from viewing an image of a man-made object such as a building containing primarily horizontal and vertical elements. A more natural scene such as foliage would result in a more circular or elliptical spectrum.

In order to return to the baseband image, the sidebands must be filtered out with a two-dimensional spatial filter. The shape of the two-dimensional frequency response shown in Figure 2.9(c) is known as a Brillouin zone.

Figure 2.9(d) shows an alternative sampling site matrix known as quincunx sampling because of the similarity to the pattern of five dots on a die. The

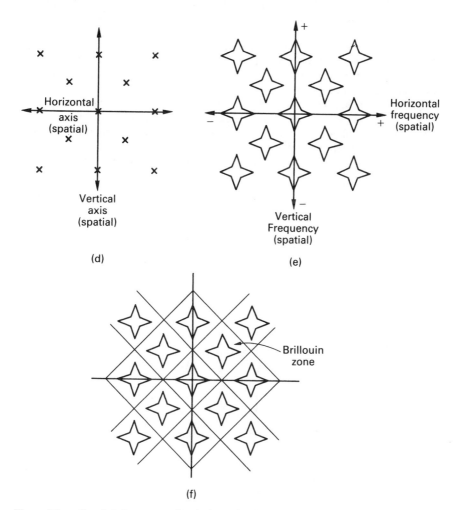

Figure 2.9 continued Quincunx sampling is shown in (d) to have a similar spectral structure (e). An appropriate Brillouin zone is required as in (f).

resultant spectrum has the same characteristic pattern as shown in Figure 2.9(e). The corresponding Brillouin zones are shown in Figure 2.9(f). Quincunx sampling offers a better compromise between diagonal and horizontal/vertical resolution but is complex to implement.

It is highly desirable to prevent spatial aliasing, since the result is visually irritating. In tube cameras the spatial aliasing will be in the vertical dimension only, since the horizontal dimension is continuously scanned. Such cameras seldom attempt to prevent vertical aliasing. CCD sensors can, however, alias in both horizontal and vertical dimensions, and so an anti-aliasing optical filter is generally fitted between the lens and the sensor. This takes the form of a plate which diffuses the image formed by the lens. Such a device can never have a sharp cut-off nor will the aperture be rectangular. The aperture of the anti-aliasing plate is in series with the aperture effect of the CCD elements, and the combination the two effectively prevents spatial aliasing, and generally gives a good balance between horizontal and vertical resolution, allowing the picture a natural appearance.

With a conventional approach, there are effectively two choices. If aliasing is permitted, the theoretical information rate of the system can be approached. If aliasing is prevented, realizable anti-aliasing filters cannot sharp cut, and the information conveyed is below system capacity.

These considerations also apply at the television display. The display must filter out spatial frequencies above one-half the sampling rate. In a conventional CRT this means that vertical optical filter should be fitted in front of the screen to render the raster invisible. Again the aperture of a simply realizable filter would attenuate too much of the wanted spectrum, and so the technique is not used.

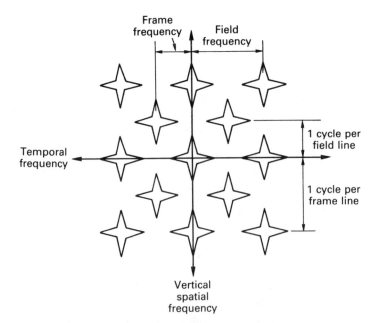

Figure 2.10 The vertical/temporal spectrum of monochrome video due to interlace.

Figure 2.10 shows the spectrum of analog monochrome video (or of an analog component). The use of interlace has a similar effect on the vertical/temporal spectrum as the use of quincunx sampling on the vertical/horizontal spectrum. The concept of the Brillouin zone cannot really be applied to reconstruction in the spatial/temporal domains. This is partly due to there being two different units in which the sampling rates are measured and partly because the temporal sampling process cannot prevent aliasing in real systems.

2.5 Motion perception

The temporal sampling rate, or field rate, of television systems is rather low at 50 or 60 Hz, and as a result temporal frequencies of more than half that will suffer aliasing. High temporal frequencies result from relative movement between the camera and objects in the field of view, as this is a form of scanning. The higher the spatial frequency of the detail in the object, and the faster the scan, the higher the temporal frequencies. It is easy to exceed the temporal frequency response of a television system even with a very slow pan, and yet the result to the viewer is not temporal aliasing. This is because the human eye will follow or track any moving object of sufficient interest. The tracking action renders the object substantially stationary with respect to the retina and so the temporal frequencies in the video spectrum due to motion are brought back to zero. Figure 2.11 illustrates the concept. In addition the temporal aperture effect of the eye is rendered immaterial and the eye then has its best spatial resolution. Clearly the eye can only track one motion at a time; the remainder of the picture will contain relative motion and will appear to the viewer to have reduced resolution due to the temporal aperture effects of the eye.

In practice the temporal aperture effect of the camera has an effect on the above process. Motion blur is proportional to the length of the temporal aperture, and shuttered CCD cameras will result in better motion portrayal than tube cameras.

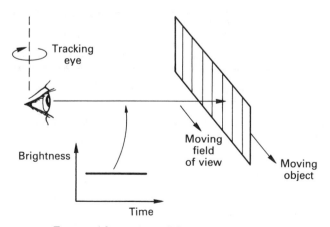

Figure 2.11 When the eye tracks motion there is no relative movement of the image on the retina.

2.6 Choice of sampling rate – component

If the reason for digitizing a video signal is simply to convey it from one place to another, then the choice of sampling frequency can be determined only by sampling theory and available filters. If, however, processing of the video in the digital domain is contemplated, the choice becomes smaller. In order to produce a two-dimensional array of samples which form rows and vertical columns, the sampling rate has to be an integer multiple of the line rate. This allows for the vertical picture processing necessary in special effects, data reduction, error concealment in recorders and standards conversion. Whilst the bandwidth needed by 525/59.94 video is less than that of 625/50, and a lower sampling rate might be used, practicality dictated that if a standard sampling rate for video components could be arrived at, then the design of standards converters would be simplified, and digital recorders would operate at a similar data rate even though the frame rates would differ in different standards. This was the goal of CCIR Recommendation 601, which combined the 625/50 input of EBU Docs Tech. 3246 and 3247 and the 525/59.94 input of SMPTE RP 125.

The result is not one sampling rate, but a family of rates based upon the magic frequency of 13.5 MHz.

Using this frequency as a sampling rate produces 858 samples in the line period of 525/59.94 and 864 samples in the line period of 625/50. For lower bandwidths, the rate can be divided by three-quarters, one-half or one-quarter to give sampling rates of 10.125, 6.75 and 3.375 MHz respectively. If the lowest frequency is considered to be 1, then the highest is 4. For maximum quality RGB working, three parallel, identical sample streams would be required, which would be denoted by 4:4:4. Colour difference signals intended for post production, where a wider colour difference bandwidth is needed, require 4:2:2 sampling for

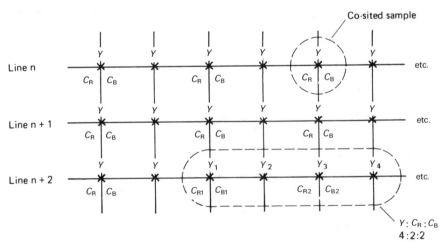

Figure 2.12 In CCIR-601 sampling mode 4:2:2, the line synchronous sampling rate of 13.5 MHz results in samples having the same position in successive lines, so that vertical columns are generated. The sampling rates of the colour difference signals C_R, C_B are one-half of that of luminance, i.e. 6.75 MHz, so that there are alternate Y-only samples and co-sited samples which describe Y, C_R and C_B. In a run of four samples, there will be four Y samples, two C_R samples and two C_B samples, hence 4:2:2.

luminance, $R-Y$ and $B-Y$ respectively; 4:2:2 has the advantage that an integer number of colour difference samples also exist in both line standards. Figure 2.12 shows the spatial arrangement given by 4:2:2 sampling. Luminance samples appear at half the spacing of colour difference samples, and half of the luminance samples are in the same physical position as a pair of colour difference samples, these being called co-sited samples. The D-1, D-5 and Digital Betacam recording formats work with 4:2:2 sampling.

Where the signal is likely to be broadcast as PAL or NTSC, a standard of 4:1:1 is acceptable, since this still delivers a colour difference bandwidth in excess of 1 MHz. Where data rate is at a premium, 3:1:1 can be used, and can still offer just about enough bandwidth for 525 lines. This would not be enough for 625 line working, but would be acceptable for ENG applications. The problem with the factors three and one is that they do not offer a columnar sampling structure, and so are not appropriate for processing systems.

Figure 2.13(a) shows the one-dimensional spectrum which results from sampling 525/59.94 video at 13.5 MHz, and Figure 2.13(b) shows the result for 625/50 video. Further details of CCIR-601 sampling can be found in Chapter 5.

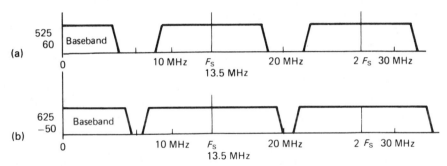

Figure 2.13 Spectra of video sampled at 13.5 MHz. In (a) the baseband 525/60 signal at left becomes the sidebands of the sampling rate and its harmonics. In (b) the same process for the 625/50 signal results in a smaller gap between baseband and sideband because of the wider bandwidth of the 625 system. The same sampling rate for both standards results in a great deal of commonality between 50 Hz and 60 Hz equipment.

The conventional TV screen has an aspect ratio of 4:3, whereas in the future an aspect ratio of 16:9 may be adopted. Expressing 4:3 as 12:9 makes it clear that the 16:9 picture is 16/12 or 4/3 times as wide. There are two ways of handling 16:9 pictures in the digital domain. One is to retain the standard sampling rate of 13.5 Mhz, which results in the horizontal resolution falling to three-quarters of its previous value; the other is to increase the sampling rate in proportion to the screen width. This results in a luminance sampling rate of 13.5 × 4/3 MHz or 18.0 MHz.

2.7 Choice of sampling rate – composite

When composite video is to be digitized, the input will be a single waveform having spectrally interleaved luminance and chroma. Any sampling rate which

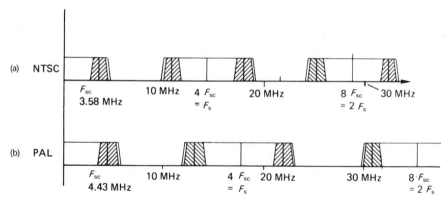

Figure 2.14 The spectra of (a) NTSC and (b) PAL where both are sampled at four times the frequency of their respective subcarriers. This high sampling rate is unnecessary to satisfy sampling theory, and so both are oversampled systems. The advantages are in the large spectral gap between baseband and sideband which allows a more gentle filter slope to be employed, and in the relative ease of colour processing at a sampling rate related to subcarrier.

allows sufficient bandwidth would convey composite video from one point to another. Indeed 13.5 MHz has been successfully used to sample PAL and NTSC. However, if simple processing in the digital domain is contemplated, there will be less choice.

In many cases it will be necessary to decode the composite signal which will require some kind of digital filter. Whilst it is possible to construct filters with any desired response, it is a fact that a digital filter whose response is simply related to the sampling rate will be much less complex to implement. This is the reasoning which has led to the near universal use of four times subcarrier sampling rate. Figure 2.14 shows the spectra of PAL and NTSC sampled at 4 × F_{sc}. It will be evident that there is a considerable space between the edge of the baseband and the lower sideband. This allows the anti-aliasing and reconstruction filters to have a more gradual cut-off, so that ripple in the passband can be reduced. This is particularly important for C-format timebase correctors and for composite digital recorders, since both are digital devices intended for use in an analog environment, and signals may have been converted to and from the digital domain many times in the course of production.

2.8 Sampling-clock jitter

The instants at which samples are taken in an ADC and the instants at which DACs make conversions must be evenly spaced, otherwise unwanted signals can be added to the video. Figure 2.15 shows the effect of sampling-clock jitter on a sloping waveform. Samples are taken at the wrong times. When these samples have passed through a system, the timebase-correction stage prior to the DAC will remove the jitter, and the result is shown in (b). The magnitude of the unwanted signal is proportional to the slope of the video waveform and so the amount of jitter which can be tolerated falls at 6 dB per octave. As the resolution of the system is increased by the use of longer sample wordlength, tolerance to jitter is further reduced. The nature of the unwanted signal depends on the spectrum of the jitter. If the jitter is random, the effect is noise-like and relatively

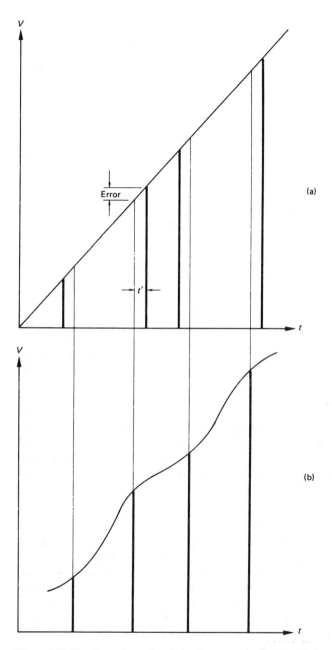

Figure 2.15 The effect of sampling timing jitter on noise. In (a) a sloping signal sampled with jitter has error proportional to the slope. When jitter is removed by reclocking, the result in (b) is noise.

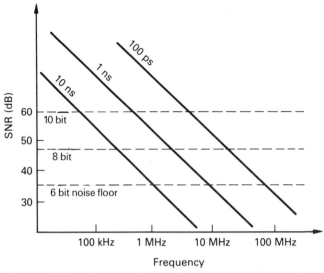

Figure 2.16 The effect of sampling-clock jitter on signal-to-noise ratio at various frequencies, compared with the theoretical noise floors with different wordlengths.

benign unless the amplitude is excessive. Figure 2.16 shows the effect of differing amounts of random jitter with respect to the noise floor of various wordlengths. Note that even small amounts of jitter can degrade a 10 bit converter to the performance of a good 8 bit unit. There is thus no point in upgrading to higher-resolution converters if the clock stability of the system is insufficient to allow their performance to be realized.

The allowable jitter is measured in picoseconds, as shown in Figure 2.15 and clearly steps must be taken to eliminate it by design. Converter clocks must be generated from clean power supplies which are well decoupled from the power used by the logic because a converter clock must have a signal-to-noise ratio of the same order as that of the signal. Otherwise noise on the clock causes jitter which in turn causes noise in the video. The same effect will be found in digital audio signals, which are perhaps more critical.

2.9 Quantizing

Quantizing is the process of expressing some infinitely variable quantity by discrete or stepped values. Quantizing turns up in a remarkable number of everyday guises. Figure 2.17 shows that an inclined ramp enables infinitely variable height to be achieved, whereas a step-ladder allows only discrete heights to be had. A step-ladder quantizes height. When accountants round off sums of money to the nearest pound or dollar they are quantizing. Time passes continuously, but the display on a digital clock changes suddenly every minute because the clock is quantizing time.

In video and audio the values to be quantized are infinitely variable voltages from an analog source. Strict quantizing is a process which operates in the voltage domain only. For the purpose of studying the quantizing of a single sample, time is assumed to stand still. This is achieved in practice either by the

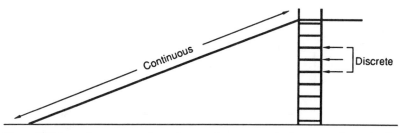

Figure 2.17 An analog parameter is continuous whereas a quantized parameter is restricted to certain values. Here the sloping side of a ramp can be used to obtain any height whereas a ladder only allows discrete heights.

use of a track-hold circuit or the adoption of a quantizer technology such as a flash converter which operates before the sampling stage.

Figure 2.18(a) shows that the process of quantizing divides the voltage range up into quantizing intervals Q, also referred to as steps S. In applications such as telephony these may advantageously be of differing size, but for digital video the quantizing intervals are made as identical as possible. If this is done, the binary numbers which result are truly proportional to the original analog voltage, and the digital equivalents of mixing and gain changing can be performed by adding and multiplying sample values. If the quantizing intervals are unequal this cannot be done. When all quantizing intervals are the same, the term uniform quantizing is used. The term linear quantizing will be found, but this is, like military intelligence, a contradiction in terms.

The term LSB (least significant bit) will also be found in place of quantizing interval in some treatments, but this is a poor term because quantizing works in the voltage domain. A bit is not a unit of voltage and can only have two values. In studying quantizing voltages within a quantizing interval will be discussed, but there is no such thing as a fraction of a bit.

Whatever the exact voltage of the input signal, the quantizer will locate the quantizing interval in which it lies. In what may be considered a separate step, the quantizing interval is then allocated a code value which is typically some form of binary number. The information sent is the number of the quantizing interval in which the input voltage lay. Whereabouts that voltage lay within the interval is not conveyed, and this mechanism puts a limit on the accuracy of the quantizer. When the number of the quantizing interval is converted back to the analog domain, it will result in a voltage at the centre of the quantizing interval as this minimizes the magnitude of the error between input and output. The number range is limited by the wordlength of the binary numbers used. In an 8 bit system, 256 different quantizing intervals exist, although in digital video the ones at the extreme ends of the range are reserved for synchronizing.

It is possible to draw a transfer function for such an ideal quantizer followed by an ideal DAC, and this is also shown in Figure 2.18. A transfer function is simply a graph of the output with respect to the input. In electronics, when the term linearity is used, this generally means the overall straightness of the transfer function. Linearity is a goal in video and audio converters, yet it will be seen that an ideal quantizer is anything but linear.

Figure 2.18(b) shows that the transfer function is somewhat like a staircase, and blanking level is half-way up a quantizing interval, or on the centre of a

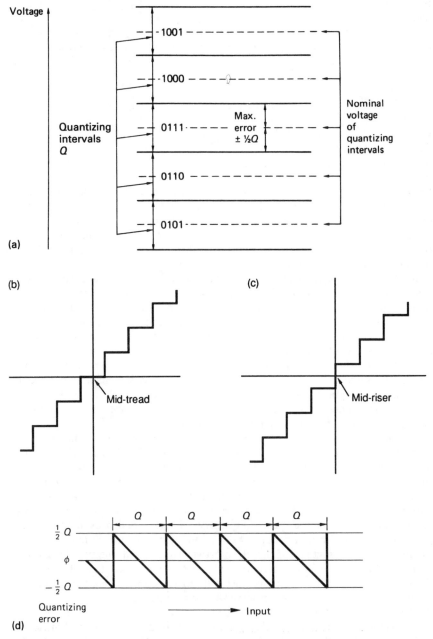

Figure 2.18 Quantizing assigns discrete numbers to variable voltages. All voltages within the same quantizing interval are assigned the same number which causes a DAC to produce the voltage at the centre of the intervals shown by the dashed lines in (a). This is the characteristic of the mid-tread quantizer shown in (b). An alternative system is the mid-riser system shown in (c). Here 0 volts analog falls between two codes and there is no code for zero. Such quantizing cannot be used prior to signal processing because the number is no longer proportional to the voltage. Quantizing error cannot exceed $\pm\frac{1}{2}Q$ as shown in (d).

tread. This is the so-called mid-tread quantizer which is universally used in video and audio.

Quantizing causes a voltage error in the sample which is given by the difference between the actual staircase transfer function and the ideal straight line. This is shown in Figure 2.18(d) to be a sawtooth-like function which is periodic in Q. The amplitude cannot exceed $\pm\frac{1}{2}Q$ peak-to-peak unless the input is so large that clipping occurs.

Quantizing error can also be studied in the time domain where it is better to avoid complicating matters with the aperture effect of the DAC. For this reason it is assumed here that output samples are of negligible duration. Then impulses from the DAC can be compared with the original analog waveform and the difference will be impulses representing the quantizing error waveform. This has been done in Figure 2.19. The horizontal lines in the drawing are the boundaries between the quantizing intervals, and the curve is the input waveform. The vertical bars are the quantized samples which reach to the centre of the quantizing interval. The quantizing error waveform shown in (b) can be thought of as an unwanted signal which the quantizing process adds to the perfect original. If a very small input signal remains within one quantizing interval, the quantizing error *is* the signal.

As the transfer function is non-linear, ideal quantizing can cause distortion. As a result practical digital video equipment deliberately uses non-ideal quantizers to achieve linearity.

As the magnitude of the quantizing error is limited, its effect can be minimized by making the signal larger. This will require more quantizing intervals and more bits to express them. The number of quantizing intervals multiplied by their size gives the quantizing range of the converter. A signal outside the range will be clipped. Provided that clipping is avoided, the larger the signal the less will be the effect of the quantizing error.

Where the input signal exercises the whole quantizing range and has a complex waveform (such as from a contrasty, detailed scene), successive samples will have widely varying numerical values and the quantizing error on a given sample will be independent of that on others. In this case the size of the quantizing error will be distributed with equal probability between the limits. Figure 2.19(c) shows the resultant uniform probability density. In this case the unwanted signal added by quantizing is an additive broadband noise uncorrelated with the signal, and it is appropriate in this case to call it quantizing noise. This is not quite the same as thermal noise which has a Gaussian (bell-shaped) probability shown in Figure 2.19(d). The difference is of no consequence as in the large-signal case the noise is masked by the signal. Under these conditions, a meaningful signal-to-noise ratio can be calculated by taking the ratio between the largest signal amplitude which can be accommodated without clipping and the error amplitude. By way of example, an 8 bit system will offer very nearly 50 dB SNR.

Whilst the above result is true for a large, complex input waveform, treatments which then assume that quantizing error is *always* noise give results which are at variance with reality. The expression above is only valid if the probability density of the quantizing error is uniform. Unfortunately at low depths of modulation, and particularly with flat fields or simple pictures, this is not the case.

At low modulation depth, quantizing error ceases to be random, and becomes a function of the input waveform and the quantizing structure as Figure 2.19

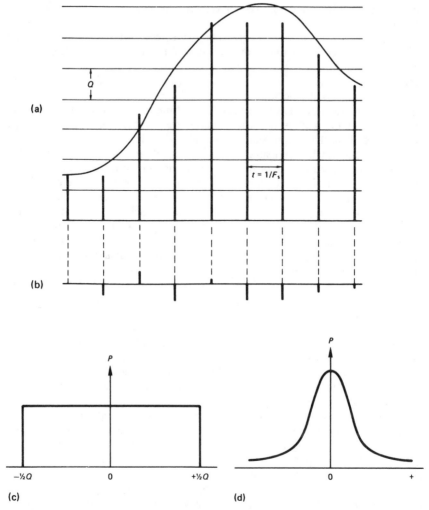

Figure 2.19 In (a) an arbitrary signal is represented to finite accuracy by PAM needles whose peaks are at the centre of the quantizing intervals. The errors caused can be thought of as an unwanted signal (b) added to the original. In (c) the amplitude of a quantizing error needle will be from $-\frac{1}{2}Q$ to $+\frac{1}{2}Q$ with equal probability. Note, however, that white noise in analog circuits generally has Gaussian amplitude distribution, shown in (d).

showed. Once an unwanted signal becomes a deterministic function of the wanted signal, it has to be classed as a distortion rather than a noise. Distortion can also be predicted from the non-linearity, or staircase nature, of the transfer function. With a large signal, there are so many steps involved that we must stand well back, and a staircase with 256 steps appears to be a slope. With a small signal there are few steps and they can no longer be ignored.

The effect can be visualized readily by considering a television camera viewing a uniformly painted wall. The geometry of the lighting and the coverage of the lens means that the brightness is not absolutely uniform, but falls slightly at the ends of the TV lines. After quantizing, the gently sloping waveform is replaced by one which stays at a constant quantizing level for many sampling periods and then suddenly jumps to the next quantizing level. The picture then consists of areas of constant brightness with steps between, resembling nothing more than a contour map; hence the use of the term *contouring* to describe the effect.

Needless to say the occurrence of contouring precludes the use of an ideal quantizer for high-quality work. There is little point in studying the adverse effects further as they should be and can be eliminated completely in practical equipment by the use of dither. The importance of correctly dithering a quantizer cannot be emphasized enough, since failure to dither irrevocably distorts the converted signal: there can be no process which will subsequently remove that distortion.

2.10 Introduction to dither

At high signal levels, quantizing error is effectively noise. As the depth of modulation falls, the quantizing error of an ideal quantizer becomes more strongly correlated with the signal and the result is distortion, visible as

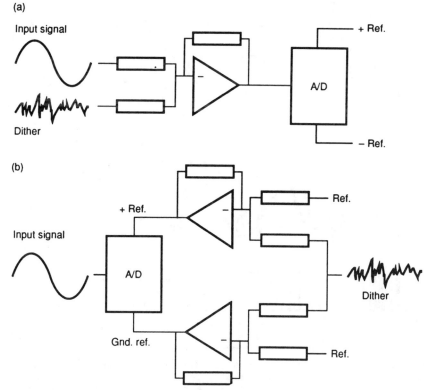

Figure 2.20 Dither can be applied to a quantizer in one of two ways. In (a) the dither is linearly added to the analog input signal, whereas in (b) it is added to the reference voltages of the quantizer.

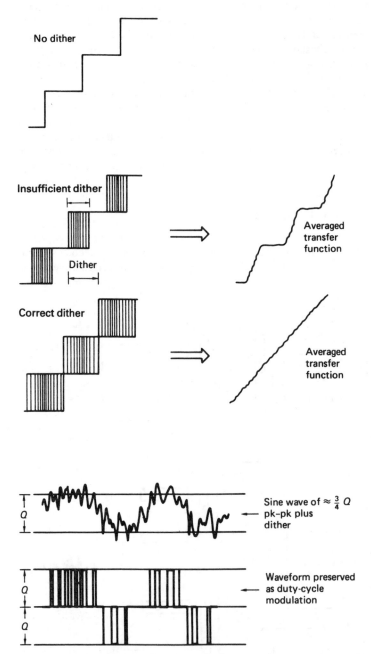

Figure 2.21 Wideband dither of the appropriate level linearizes the transfer function to produce noise instead of distortion. This can be confirmed by spectral analysis. In the voltage domain, dither causes frequent switching between codes and preserves resolution in the duty cycle of the switching.

contouring. If the quantizing error can be decorrelated from the input in some way, the system can remain linear but noisy. Dither performs the job of decorrelation by making the action of the quantizer unpredictable and gives the system a noise floor like an analog system.[4,5]

All practical digital video systems use non-subtractive dither where the dither signal is added prior to quantization and no attempt is made to remove it at the DAC.[6] The introduction of dither prior to a conventional quantizer inevitably causes a slight reduction in the signal-to-noise ratio attainable, but this reduction is a small price to pay for the elimination of non-linearities.

The ideal (noiseless) quantizer of Figure 2.18 has fixed quantizing intervals and must always produce the same quantizing error from the same signal. In Figure 2.20 it can be seen that an ideal quantizer can be dithered by linearly adding a controlled level of noise either to the input signal or to the reference voltage which is used to derive the quantizing intervals. There are several ways of considering how dither works, all of which are equally valid.

The addition of dither means that successive samples effectively find the quantizing intervals in different places on the voltage scale. The quantizing error becomes a function of the dither, rather than a predictable function of the input signal. The quantizing error is not eliminated, but the subjectively unacceptable distortion is converted into a broadband noise which is more benign to the viewer.

Some alternative ways of looking at dither are shown in Figure 2.21. Consider the situation where a low-level input signal is changing slowly within a quantizing interval. Without dither, the same numerical code is output for a number of sample periods, and the variations within the interval are lost. Dither has the effect of forcing the quantizer to switch between two or more states. The higher the voltage of the input signal within a given interval, the more probable it becomes that the output code will take on the next higher value. The lower the input voltage within the interval, the more probable it is that the output code will take the next lower value. The dither has resulted in a form of duty cycle modulation, and the resolution of the system has been extended indefinitely instead of being limited by the size of the steps.

Dither can also be understood by considering what it does to the transfer function of the quantizer. This is normally a perfect staircase, but in the presence of dither it is smeared horizontally until with a certain amplitude the average transfer function becomes straight.

2.11 Requantizing and digital dither

Recent ADC technology allows the resolution of video samples to be raised from 8 bits to 10 or even 12 bits. The situation then arises that an existing 8 bit device such as a digital VTR needs to be connected to the output of an ADC with greater wordlength. The words need to be shortened in some way.

When a sample value is attenuated, the extra low-order bits which come into existence below the radix point preserve the resolution of the signal and the dither in the least significant bit(s) which linearizes the system. The same word extension will occur in any process involving multiplication, such as digital filtering. It will subsequently be necessary to shorten the wordlength. Low-order bits must be removed in order to reduce the resolution whilst keeping the signal

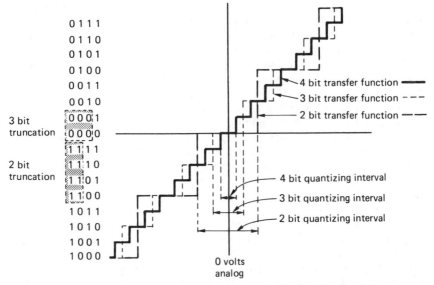

Figure 2.22 Shortening the wordlength of a sample reduces the number of codes which can describe the voltage of the waveform. This makes the quantizing steps bigger hence the term requantizing. It can be seen that simple truncation or omission of the bits does not give analogous behaviour. Rounding is necessary to give the same result as if the larger steps had been used in the original conversion.

magnitude the same. Even if the original conversion was correctly dithered, the random element in the low-order bits will now be some way below the end of the intended word. If the word is simply truncated by discarding the unwanted low-order bits or rounded to the nearest integer the linearizing effect of the original dither will be lost.

Shortening the wordlength of a sample reduces the number of quantizing intervals available without changing the signal amplitude. As Figure 2.22 shows, the quantizing intervals become larger and the original signal is *requantized* with the new interval structure. This will introduce requantizing distortion having the same characteristics as quantizing distortion in an ADC. It then is obvious that when shortening the wordlength of a 10 bit converter to 8 bits, the two low-order bits must be removed in a way that displays the same overall quantizing structure as if the original converter had been only of 8 bit wordlength. It will be seen from Figure 2.22 that truncation cannot be used because it does not meet the above requirement but results in signal-dependent offsets because it always rounds in the same direction. Proper numerical rounding is essential in video applications because it accurately simulates analog quantizing to the new interval size. Unfortunately the 10 bit converter will have a dither amplitude appropriate to quantizing intervals one-quarter the size of an 8 bit unit and the result will be highly non-linear.

In practice, the wordlength of samples must be shortened in such a way that the requantizing error is converted to noise rather than distortion. One technique which meets this requirement is to use digital dithering[7] prior to rounding. This is directly equivalent to the analog dithering in an ADC.

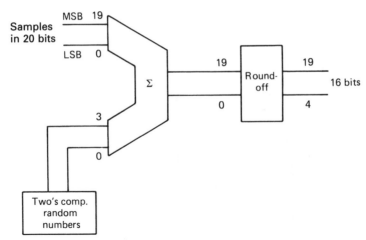

Figure 2.23 In a simple digital dithering system, two's complement values from a random number generator are added to low-order bits of the input. The dithered values are then rounded up or down according to the value of the bits to be removed. The dither linearizes the requantizing.

Digital dither is a pseudo-random sequence of numbers. If it is required to simulate the analog dither signal of Figures 2.20 and 2.21, then it is obvious that the noise must be bipolar so that it can have an average voltage of zero. Two's complement coding must be used for the dither values as it is for the audio samples.

Figure 2.23 shows a simple digital dithering system (i.e. one without noise shaping) for shortening sample wordlength. The output of a two's complement pseudo-random sequence generator of appropriate wordlength is added to input samples prior to rounding. The most significant of the bits to be discarded is examined in order to determine whether the bits to be removed sum to more or less than half a quantizing interval. The dithered sample is either rounded down, i.e. the unwanted bits are simply discarded, or rounded up, i.e. the unwanted bits are discarded but 1 is added to the value of the new short word. The rounding process is no longer deterministic because of the added dither which provides a linearizing random component.

If this process is compared with that of Figure 2.20 it will be seen that the principles of analog and digital dither are identical; the processes simply take place in different domains using two's complement numbers which are rounded or voltages which are quantized as appropriate. In fact quantization of an analog dithered waveform is identical to the hypothetical case of rounding after bipolar digital dither where the number of bits to be removed is infinite, and remains identical for practical purposes when as few as eight bits are to be removed. Analog dither may actually be generated from bipolar digital dither (which is no more than random numbers with certain properties) using a DAC.

2.12 Basic digital-to-analog conversion

This direction of conversion will be discussed first, since ADCs often use embedded DACs in feedback loops.

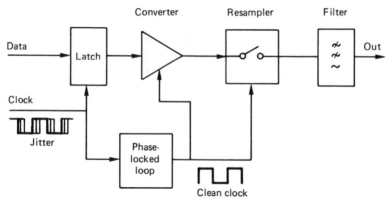

Figure 2.24 The components of a conventional converter. A jitter-free clock drives the voltage conversion, whose output may be resampled prior to reconstruction.

The purpose of a digital-to-analog converter is to take numerical values and reproduce the continuous waveform that they represent. Figure 2.24 shows the major elements of a conventional conversion subsystem, i.e. one in which oversampling is not employed. The jitter in the clock needs to be removed with a VCO or VCXO. Sample values are buffered in a latch and fed to the converter element which operates on each cycle of the clean clock. The output is then a voltage proportional to the number for at least a part of the sample period. A resampling stage may be found next, in order to remove switching transients, reduce the aperture ratio or allow the use of a converter which takes a substantial part of the sample period to operate. The resampled waveform is then presented to a reconstruction filter which rejects frequencies above the audio band.

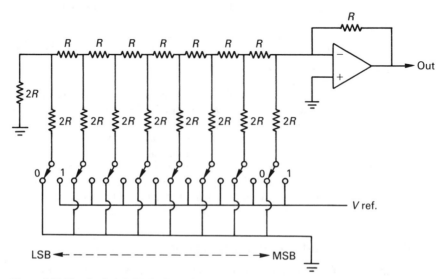

Figure 2.25 The classical R–2R DAC requires precise resistance values and 'perfect' switches.

This section is primarily concerned with the implementation of the converter element. The most common way of achieving this conversion is to control binary-weighted currents and sum them in a virtual earth. Figure 2.25 shows the classical $R-2R$ DAC structure. This is relatively simple to construct, but the resistors have to be extremely accurate. To see why this is so, consider the example of Figure 2.26. In (a) the binary code is about to have a major overflow, and all the low-order currents are flowing. In (b), the binary input has increased

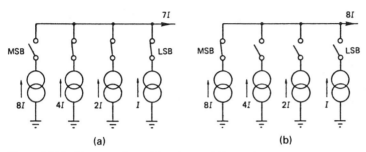

(a)   (b)

Figure 2.26 In (a) current flow with an input of 0111 is shown. In (b) current flow with input code one greater is shown.

by 1, and only the most significant current flows. This current must equal the sum of all the others plus 1. The accuracy must be such that the step size is within the required limits. In this 8 bit example, if the step size needs to be a rather casual 10% accurate, the necessary accuracy is only one part in 2560, but for a 10 bit system it would become one part in 10 240. This degree of accuracy is difficult to achieve and maintain in the presence of ageing and temperature change.

2.13 Basic analog-to-digital conversion

The general principle of a quantizer is that different quantized voltages are compared with the unknown analog input until the closest quantized voltage is found. The code corresponding to this becomes the output. The comparisons can be made in turn with the minimal amount of hardware, or simultaneously with more hardware.

The flash converter is probably the simplest technique available for PCM video conversion. The principle is shown in Figure 2.27. The threshold voltage of every quantizing interval is provided by a resistor chain which is fed by a reference voltage. This reference voltage can be varied to determine the sensitivity of the input. There is one voltage comparator connected to every reference voltage, and the other input of all of the comparators is connected to the analog input. A comparator can be considered to be a 1 bit ADC. The input voltage determines how many of the comparators will have a true output. As one comparator is necessary for each quantizing interval, then, for example, in an 8 bit system there will be 255 binary comparator outputs, and it is necessary to use a priority encoder to convert these to a binary code. Note that the quantizing stage is asynchronous; comparators change state as and when the variations in the input waveform result in a reference voltage being crossed. Sampling takes place when the comparator outputs are clocked into a subsequent latch. This is an example of quantizing before sampling

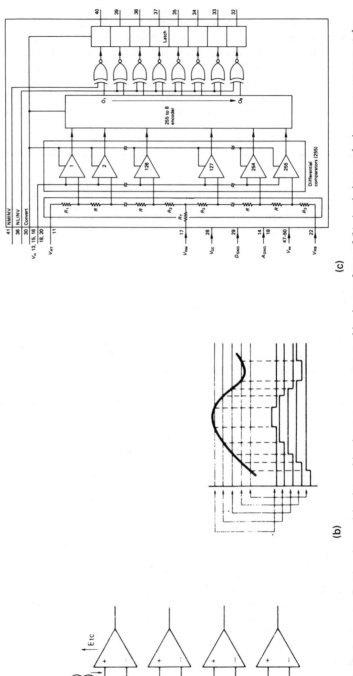

Figure 2.27 The flash converter. In (a) each quantizing interval has its own comparator, resulting in waveforms of (b). A priority encoder is necessary to convert the comparator outputs to a binary code. Shown in (c) is a typical 8 bit flash converter primarily intended for video applications. (Courtesy TRW).

as was illustrated in Figure 2.1. Although the device is simple in principle, it contains a lot of circuitry and can only be practicably implemented on a chip. The analog signal has to drive a lot of inputs, which results in a significant parallel capacitance, and a low-impedance driver is essential to avoid restricting the slewing rate of the input. The extreme speed of a flash converter is a distinct advantage in oversampling. Because computation of all bits is performed simultaneously, no track-hold circuit is required, and droop is eliminated. Figure 2.27(c) shows a flash converter chip. Note the resistor ladder and the comparators followed by the priority encoder. The MSB can be selectively inverted so that the device can be used either in offset binary or two's complement mode.

The flash converter is ubiquitous in digital video because of the high speed necessary. For audio purposes, longer wordlengths are used and different techniques will be necessary.

2.14 Oversampling

Oversampling means using a sampling rate which is greater (generally substantially greater) than the Nyquist rate. Neither sampling theory nor quantizing theory *requires* oversampling to be used to obtain a given signal quality, but Nyquist rate conversion places extremely high demands on component accuracy when a converter is implemented. Oversampling allows a given signal quality to be reached without requiring very close tolerance, and therefore expensive, components.

Figure 2.28 shows the main advantages of oversampling. In (a) it will be seen that the use of a sampling rate considerably above the Nyquist rate allows the

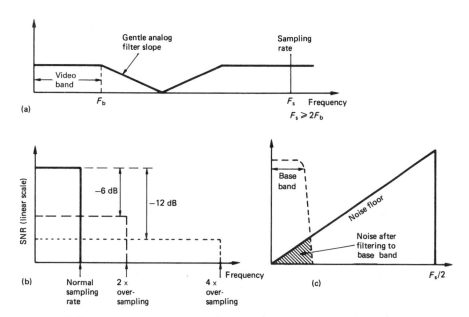

Figure 2.28 Oversampling has a number of advantages. In (a) it allows the slope of analog filters to be relaxed. In (b) it allows the resolution of converters to be extended. In (c) a *noise-shaped* converter allows a disproportionate improvement in resolution.

anti-aliasing and reconstruction filters to be realized with a much more gentle cut-off slope. There is then less likelihood of phase linearity and ripple problems in the passband.

Figure 2.28(b) shows that information in an analog signal is two-dimensional and can be depicted as an area which is the product of bandwidth and the linearly expressed signal-to-noise ratio. The figure also shows that the same amount of information can be conveyed down a channel with a SNR of half as much (6 dB less) if the bandwidth used is doubled, with 12 dB less SNR if bandwidth is quadrupled, and so on, provided that the modulation scheme used is perfect.

The information in an analog signal can be conveyed using some analog modulation scheme in any combination of bandwidth and SNR which yields the appropriate channel capacity. If bandwidth is replaced by sampling rate and SNR is replaced by a function of wordlength, the same must be true for a digital signal as it is no more than a numerical analog. Thus raising the sampling rate potentially allows the wordlength of each sample to be reduced without information loss.

Information theory predicts that if a signal is spread over a much wider bandwidth by some modulation technique, the SNR of the demodulated signal can be higher than that of the channel it passes through, and this is also the case in digital systems. The concept is illustrated in Figure 2.29. In (a) 4 bit samples are delivered at sampling rate F. As 4 bits have 16 combinations, the information rate

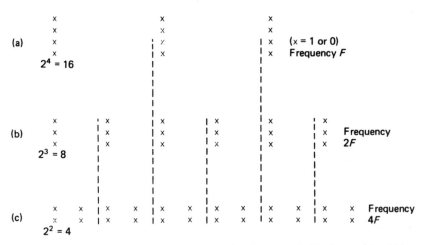

Figure 2.29 Information rate can be held constant when frequency doubles by removing 1 bit from each word. In all cases here it is 16F. Note bit rate of (c) is double that of (a). Data storage in oversampled form is inefficient.

is 16F. In (b) the same information rate is obtained with 3 bit samples by raising the sampling rate to 2F and in (c) 2 bit samples having four combinations require to be delivered at a rate of 4F. Whilst the information rate has been maintained, it will be noticed that the bit rate of (c) is twice that of (a). The reason for this is shown in Figure 2.30. A single binary digit can only have two states; thus it can only convey two pieces of information, perhaps 'yes' or 'no'. Two binary digits together can have four states, and can thus convey four pieces of information,

	0 = No 1 = Yes	00 = Spring 01 = Summer 10 = Autumn 11 = Winter	000 do 001 re 010 mi 011 fa 100 so 101 la 110 te 111 do	0000 0 0001 1 0010 2 0011 3 0100 4 0101 5 0110 6 0111 7 1000 8 1001 9 1010 A 1011 B 1100 C 1101 D 1110 E 1111 F		000 3FF	Digital video sample values
No of bits	1	2	3	4		10	
Information per word	2	4	8	16		1024	
Information per bit	2	2	≈3	4		102	

Figure 2.30 The amount of information per bit increases disproportionately as wordlength increases. It is always more efficient to use the longest words possible at the lowest word rate. It will be evident that 10 bit PCM is one hundred times as efficient as delta modulation. Oversampled data are also inefficient for storage.

perhaps 'spring summer autumn or winter', which is two pieces of information per bit. Three binary digits grouped together can have eight combinations, and convey eight pieces of information, perhaps 'doh re mi fah so lah te or doh', which is nearly three pieces of information per digit. Clearly the further this principle is taken, the greater the benefit. In a 10 bit system, each bit is worth one hundred pieces of information. It is always more efficient, in information-capacity terms, to use the combinations of long binary words than to send single bits for every piece of information. The greatest efficiency is reached when the longest words are sent at the slowest rate which must be the Nyquist rate. This is one reason why PCM recording is more common than delta modulation, despite the simplicity of implementation of the latter type of converter. PCM simply makes more efficient use of the capacity of the binary channel.

As a result, oversampling is confined to converter technology where it gives specific advantages in implementation. The storage or transmission system will usually employ PCM, where the sampling rate is a little more than twice the input bandwidth. Figure 2.31 shows a digital VTR using oversampling converters. The ADC runs at n times the Nyquist rate, but once in the digital domain the rate needs to be reduced in a type of digital filter called a *decimator*. The output of this is conventional Nyquist rate PCM, according to the tape format, which is then recorded. On replay the sampling rate is raised once more in a further type of digital filter called an *interpolator*. The system now has the best of both worlds: using oversampling in the converters overcomes the shortcomings of analog anti-aliasing and reconstruction filters and the wordlength of the converter

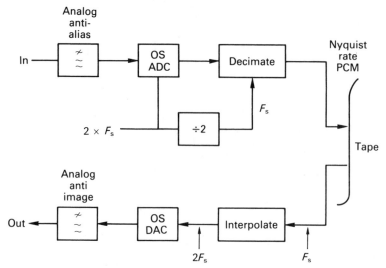

Figure 2.31 An oversampling DVTR. The converters run faster than sampling theory suggests to ease analog filter design. Sampling-rate reduction allows efficient PCM recording on tape.

elements is reduced making them easier to construct; the recording is made with Nyquist-rate PCM which minimizes tape consumption.

Oversampling is a method of overcoming practical implementation problems by replacing a single critical element or bottleneck by a number of elements whose overall performance is what counts. As Hauser[8] properly observed, oversampling tends to overlap the operations which are quite distinct in a conventional converter. In earlier sections of this chapter, the vital subjects of filtering, sampling, quantizing and dither have been treated almost independently. Figure 2.32(a) shows that it is possible to construct an ADC of predictable performance by a taking a suitable anti-aliasing filter, a sampler, a dither source and a quantizer and assembling them like building bricks. The bricks are effectively in series and so the performance of each stage can only limit the overall performance. In contrast Figure 2.32(b) shows that with oversampling the overlap of operations allows different processes to augment one another allowing a synergy which is absent in the conventional approach.

If the oversampling factor is n, the analog input must be bandwidth limited to $nF_s/2$ by the analog anti-aliasing filter. This unit need only have a flat frequency response and phase linearity within the video band. Analog dither of an amplitude compatible with the quantizing interval size is added prior to sampling at nF_s and quantizing.

Next, the anti-aliasing function is completed in the digital domain by a low-pass filter which cuts off at $F_s/2$. Using an appropriate architecture this filter can be absolutely phase linear and implemented to arbitrary accuracy. The filter can be considered to be the demodulator of Figure 2.28 where the SNR improves as the bandwidth is reduced. The wordlength can be expected to increase. The multiplications taking place within the filter extend the wordlength considerably more than the bandwidth reduction alone would indicate. The analog filter serves

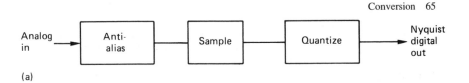

(a)

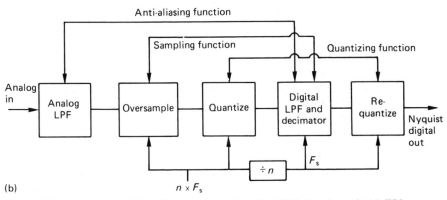

(b)

Figure 2.32 A conventional ADC performs each step in an identifiable location as in (a). With oversampling, many of the steps are distributed as shown in (b).

only to prevent aliasing into the baseband at the oversampling rate; the signal spectrum is determined with greater precision by the digital filter.

With the information spectrum now Nyquist limited, the sampling process is completed when the rate is reduced in the decimator. One sample in n is retained.

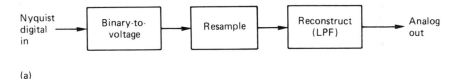

(a)

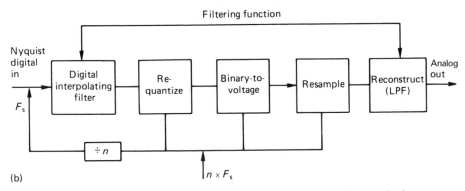

(b)

Figure 2.33 A conventional DAC in (a) is compared with the oversampling implementation in (b).

The excess wordlength extension due to the anti-aliasing filter arithmetic must then be removed. Digital dither is added, completing the dither process, and the quantizing process is completed by requantizing the dithered samples to the appropriate wordlength which will be greater than the wordlength of the first quantizer. Alternatively noise shaping may be employed.

Figure 2.33(a) shows the building brick approach of a conventional DAC. The Nyquist rate samples are converted to analog voltages and then a steep-cut analog low-pass filter is needed to reject the sidebands of the sampled spectrum.

Figure 2.33(b) shows the oversampling approach. The sampling rate is raised in an interpolator which contains a low-pass filter which restricts the baseband spectrum to the audio bandwidth shown. A large frequency gap now exists between the baseband and the lower sideband. The multiplications in the interpolator extend the wordlength considerably and this must be reduced within the capacity of the DAC element by the addition of digital dither prior to requantizing.

Oversampling may also be used to considerable benefit in other dimensions. Figure 2.34 shows how vertical oversampling can be used to increase the resolution of a TV system. A 1250 line camera is used as the input device, but the 1250 line signal is fed to a standards converter which reduces the number of lines to 625. The standards converter must incorporate a vertical low-pass spatial filter to prevent aliasing when the vertical sampling rate is effectively halved. As this will be a digital filter, it can have arbitrarily accurate peformance, including a flat passband and steep cut-off slope. The combination of the vertical aperture

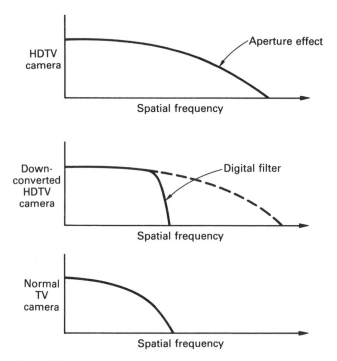

Figure 2.34 Using an HDTV camera with down conversion is a form of oversampling and gives better results than a normal camera because the aperture effect is overcome.

effect of the 1250 line camera and the vertical LPF in the standards converter gives a better spatial frequency response than could be achieved with a 625 line camera. The improvement in subjective quality is quite noticeable in practice.

In the case of display technology, oversampling can also be used, this time to render the raster invisible and to improve the vertical aperture of the display. Once more a standards converter is required, but this now doubles the number of input lines using interpolation. Again the filter can have arbitrary accuracy. The vertical aperture of the 1250 line display does not affect the passband of the input signal because of the use of oversampling.

Oversampling can also be used in the time domain in order to reduce or eliminate display flicker. A different type of standards converter is necessary which doubles the input field rate by interpolation. The standards converter must use motion compensation otherwise moving objects will not be correctly positioned in intermediate fields and will suffer from judder. Motion compensation is considered in Chapter 3.

References

1. SHANNON, C.E., A mathematical theory of communication. *Bell Syst. Tech. J.*, **27**, 379 (1948)
2. JERRI, A.J., The Shannon sampling theorem – its various extensions and applications: a tutorial review. *Proc. IEEE*, **65**, 1565–1596 (1977)
3. KELL, R., BEDFORD, A. and TRAINER, M.A. An experimental television system, Part 2. *Proc. IRE*, **22**, 1246–1265 (1934)
4. GOODALL, W. M., Television by pulse code modulation. *Bell Syst. Tech. J.*, **30**, 33–49 (1951)
5. ROBERTS, L. G., Picture coding using pseudo-random noise. *IRE Trans. Inf. Theory*, **IT-8**, 145–154 (1962)
6. VANDERKOOY, J. and LIPSHITZ, S.P., Resolution below the least significant bit in digital systems with dither. *J. Audio Eng. Soc.*, **32**, 106–113 (1984)
7. VANDERKOOY, J. and LIPSHITZ, S.P., Digital dither. Presented at 81st Audio Engineering Society Convention (Los Angeles, 1986), preprint 2412 (C-8)
8. HAUSER, M.W., Principles of oversampling A/D conversion. *J. Audio Eng. Soc.*, **39**, 3–26 (1991)

Chapter 3

Digital processing

The conversion process expresses the analog input as a numerical code. The choice of code for video will be shown to be governed by the requirements of the processing. Within the digital domain, all signal processing must be performed by arithmetic manipulation of the code values in suitable logic circuits. All of the necessary principles of binary arithmetic and their implementation in logic are introduced here.

3.1 Pure binary code

For digital video use, the prime purpose of binary numbers is to express the values of the samples which represent the original analog video waveform. Figure 3.1 shows some binary numbers and their equivalent in decimal. The radix point has the same significance in binary: symbols to the right of it represent one half, one quarter and so on. Binary is convenient for electronic circuits, which do not get tired, but numbers expressed in binary become very long, and writing them is tedious and error-prone. The octal and hexadecimal notations are both used for writing binary since conversion is so simple. Figure 3.1 also shows that a binary number is split into groups of three or four digits starting at the least significant end, and the groups are individually converted to octal or hexadecimal digits. Since sixteen different symbols are required in hex, the letters A–F are used for the numbers above nine.

There will be a fixed number of bits in a PCM video sample, and this number determines the size of the quantizing range. In the 8 bit samples used in much digital video equipment, there are 256 different numbers. Each number represents a different analog signal voltage, and care must be taken during conversion to ensure that the signal does not go outside the converter range, or it will be clipped. In Figure 3.2(a) it will be seen that in an 8 bit pure binary system, the number range goes from 00 hex, which represents the smallest voltage, through to FF hex, which represents the largest positive voltage. The video waveform must be accommodated within this voltage range, and Figure 3.2(b) shows how this can be done for a PAL composite signal. A luminance signal is shown in Figure 3.2(c). As component digital systems only handle the active line, the quantizing range is optimized to suit the gamut of the unblanked luminance. There is a small offset in order to handle slightly misadjusted inputs.

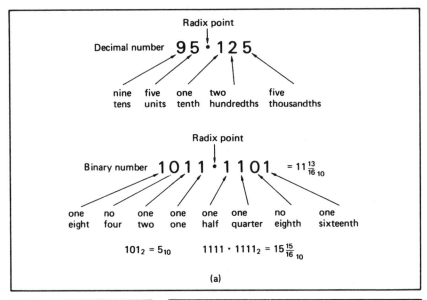

(a)

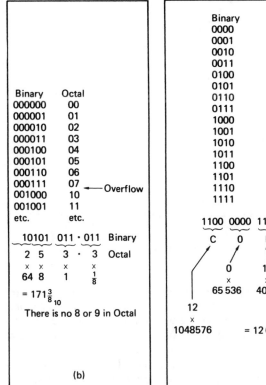

(b)

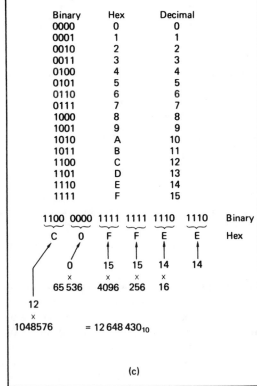

(c)

Figure 3.1 (a) Binary and decimal. (b) In octal, groups of 3 bits make one symbol 0–7. (c) In hex, groups of 4 bits make one symbol 0–F. Note how much shorter the number is in hex.

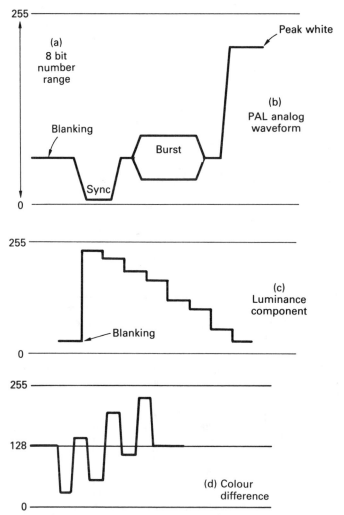

Figure 3.2 The unipolar quantizing range of an 8 bit pure binary system is shown in (a). The analog input must be shifted to fit into the quantizing range, as shown for PAL in (b). In component digital systems, sync pulses are not digitized, so the quantizing intervals can be smaller as in (c). An offset of half scale is used for colour difference signals (d).

Colour difference signals are bipolar and so blanking is in the centre of the signal range. In order to accommodate colour difference signals in the quantizing range, the blanking voltage level of the analog waveform has been shifted as in Figure 3.2(d) so that the positive and negative voltages in a real video signal can be expressed by binary numbers which are only positive. This approach is called offset binary. Strictly speaking both the composite and luminance signals are also offset binary because the blanking level is part way up the quantizing scale.

Offset binary is perfectly acceptable where the signal has been digitized only for recording or transmission from one place to another, after which it will be converted directly back to analog. Under these conditions it is not necessary for

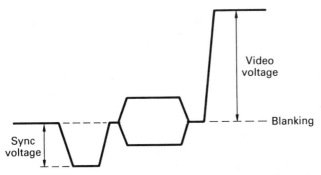

Figure 3.3 All video signal voltages are referred to blanking and must be added with respect to that level.

the quantizing steps to be uniform, provided both ADC and DAC are constructed to the same standard. In practice, it is the requirements of signal processing in the digital domain which make both non-uniform quantizing and offset binary unsuitable.

Figure 3.3 shows that analog video signal voltages are referred to blanking. The level of the signal is measured by how far the waveform deviates from blanking, and attenuation, gain and mixing all take place around blanking level. Digital vision mixing is achieved by adding sample values from two or more different sources, but unless all of the quantizing intervals are of the same size and there is no offset, the sum of two sample values will not represent the sum of the two original analog voltages. Thus sample values which have been obtained by non-uniform or offset quantizing cannot readily be processed because the binary numbers are not proportional to the signal voltage.

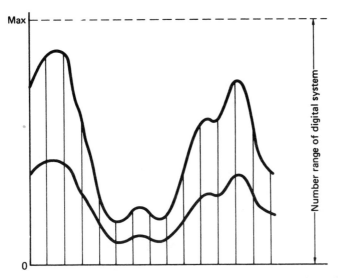

Figure 3.4 The result of an attempted attenuation in pure binary code is an offset. Pure binary cannot be used for digital video processing.

If two offset binary sample streams are added together in an attempt to perform digital mixing, the result will be that the offsets are also added and this may lead to an overflow. Similarly, if an attempt is made to attenuate by, say, 6.02 dB by dividing all of the sample values by two, Figure 3.4 shows that the offset is also divided and the waveform suffers a shifted baseline. This problem can be overcome with digital luminance signals simply by subtracting the offset from each sample before processing as this results in numbers truly proportional to the luminance voltage. This approach is not suitable for colour difference or composite signals because negative numbers would result when the analog voltage goes below blanking and pure binary coding cannot handle them. The problem with offset binary is that it works with reference to one end of the range. What is needed is a numbering system which operates symmetrically with reference to the centre of the range.

3.2 Two's complement

In the two's complement system, the upper half of the pure binary number range has been redefined to represent negative quantities. If a pure binary counter is constantly incremented and allowed to overflow, it will produce all the numbers in the range permitted by the number of available bits, and these are shown for a 4 bit example drawn around the circle in Figure 3.5. As a circle has no real

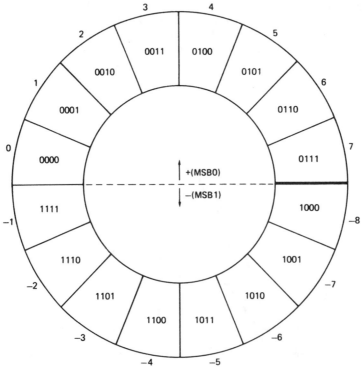

Figure 3.5 In this example of a 4 bit two's complement code, the number range is from −8 to +7. Note that the MSB determines polarity.

beginning, it is possible to consider it to start wherever it is convenient. In two's complement, the quantizing range represented by the circle of numbers does not start at zero, but starts on the diametrically opposite side of the circle. Zero is midrange, and all numbers with the MSB (most significant bit) set are considered negative. The MSB is thus the equivalent of a sign bit where 1 = minus. Two's complement notation differs from pure binary in that the most significant bit is inverted in order to achieve the half circle rotation.

Figure 3.6 shows how a real ADC is configured to produce two's complement output. In (a) an analog offset voltage equal to one-half the quantizing range is added to the bipolar analog signal in order to make it unipolar as in (b). The ADC produces positive only numbers in (c) which are proportional to the input voltage. The MSB is then inverted in (d) so that the all-zeros code moves to the centre of the quantizing range. The analog offset is often incorporated in the ADC

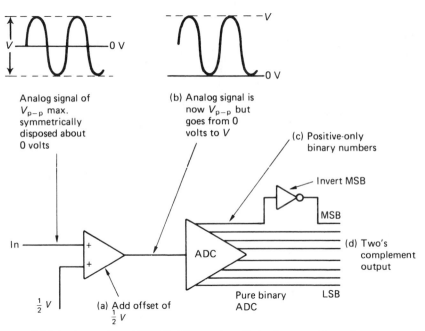

Figure 3.6 A two's complement ADC. In (a) an analog offset voltage equal to one-half the quantizing range is added to the bipolar analog signal in order to make it unipolar as in (b). The ADC produces positive only numbers in (c), but the MSB is then inverted in (d) to give a two's complement output.

as is the MSB inversion. Some converters are designed to be used in either pure binary or two's complement mode. In this case the designer must arrange the appropriate DC conditions at the input. The MSB inversion may be selectable by an external logic level. In the digital video interface standards the colour difference signals use offset binary because the codes of all zeros and all ones are at the end of the range and can be reserved for synchronizing. A digital vision mixer simply inverts the MSB of each colour difference sample to convert it to two's complement.

The two's complement system allows two sample values to be added, or mixed in video parlance, and the result will be referred to the system midrange; this is analogous to adding analog signals in an operational amplifier.

Figure 3.7 illustrates how adding two's complement samples simulates a bipolar mixing process. The waveform of input A is depicted by solid black samples, and that of B by samples with a solid outline. The result of mixing is the linear sum of the two waveforms obtained by adding pairs of sample values. The dashed lines depict the output values. Beneath each set of samples is the

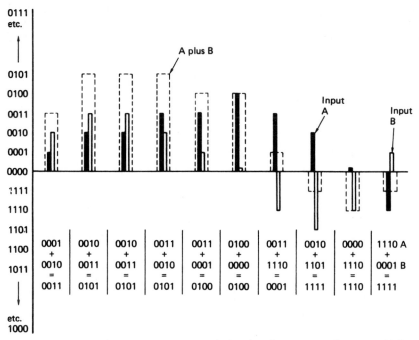

Figure 3.7 Using two's complement arithmetic, single values from two waveforms are added together with respect to midrange to give a correct mixing function.

calculation which will be seen to give the correct result. Note that the calculations are pure binary. No special arithmetic is needed to handle two's complement numbers.

It is sometimes necessary to phase-reverse or invert a digital signal. The process of inversion in two's complement is simple. All bits of the sample value are inverted to form the one's complement, and one is added. This can be checked by mentally inverting some of the values in Figure 3.5. The inversion is transparent and performing a second inversion gives the original sample values.

Using inversion, signal subtraction can be performed using only adding logic. The inverted input is added to perform a subtraction, just as in the analog domain. This permits a significant saving in hardware complexity, since only carry logic is necessary and no borrow mechanism need be supported.

In summary, two's complement notation is the most appropriate scheme for bipolar signals, and allows simple mixing in conventional binary adders. It is in virtually universal use in digital video and audio processing.

Two's complement numbers can have a radix point and bits below it just as pure binary numbers can. It should, however, be noted that in two's complement, if a radix point exists, numbers to the right of it are added. For example 1100.1 is not −4.5, it is −4 + 0.5 = −3.5.

3.3 Introduction to digital logic

However complex a digital process, it can be broken down into smaller stages until finally one finds that there are really only two basic types of element in use, and these can be combined in some way and supplied with a clock to implement virtually any process. Figure 3.8 shows that the first type is a *logical* element. This produces an output which is a logical function of the input with minimal delay. The second type is a *storage* element which samples the state of the input(s) when clocked and holds or delays that state. The strength of binary logic

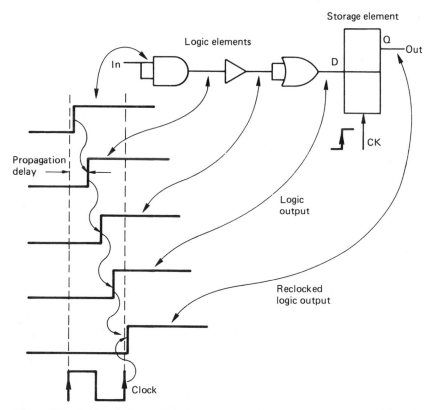

Figure 3.8 Logic elements have a finite propagation delay between input and output and cascading them delays the signal an arbitrary amount. Storage elements sample the input on a clock edge and can return a signal to near coincidence with the system clock. This is known as reclocking. Reclocking eliminates variations in propagation delay in logic elements.

is that the signal has only two states, and considerable noise and distortion of the binary waveform can be tolerated before the state becomes uncertain. At every logical element, the signal is compared with a threshold, and can thus can pass through any number of stages without being degraded. In addition, the use of a storage element at regular locations throughout logic circuits eliminates time variations or jitter. Figure 3.8 shows that if the inputs to a logic element change, the output will not change until the *propagation delay* of the element has elapsed. However, if the output of the logic element forms the input to a storage element, the output of that element will not change until the input is sampled *at the next clock edge*. In this way the signal edge is aligned to the system clock and the propagation delay of the logic becomes irrelevant. The process is known as reclocking.

The two states of the signal when measured with an oscilloscope are simply two voltages, usually referred to as high and low. As there are only two states, there can only be *true* or *false* meanings. The true state of the signal can be assigned by the designer to either voltage state. When a high voltage represents a true logic condition and a low voltage represents a false condition, the system is known as *positive logic*, or *high true* logic. This is the usual system, but sometimes the low voltage represents the true condition and the high voltage represents the false condition. This is known as *negative logic* or *low true* logic. Provided that everyone is aware of the logic convention in use, both work equally well.

In logic systems, all logical functions, however complex, can be configured from combinations of a few fundamental logic elements or *gates*. It is not profitable to spend too much time debating which are the truly fundamental ones, since most can be made from combinations of others. Figure 3.9 shows the important simple gates and their derivatives, and introduces the logical expressions to describe them, which can be compared with the truth-table notation. The figure also shows the important fact that when negative logic is used, the OR gate function interchanges with that of the AND gate.

If numerical quantities need to be conveyed down the two-state signal paths described here, then the only appropriate numbering system is binary, which has only two symbols, 0 and 1. Just as positive or negative logic could be used for the truth of a logical binary signal, it can also be used for a numerical binary signal. Normally, a high voltage level will represent a binary 1 and a low voltage will represent a binary 0, described as a 'high for a one' system. Clearly a 'low for a one' system is just as feasible. Decimal numbers have several columns, each of which represents a different power of ten; in binary the column position specifies the power of two.

Several binary digits or bits are needed to express the value of a binary video sample. These bits can be conveyed at the same time by several signals to form a parallel system, which is most convenient inside equipment or for short distances because it is inexpensive, or one at a time down a single signal path, which is more complex, but convenient for cables between pieces of equipment because the connectors require fewer pins. When a binary system is used to convey numbers in this way, it can be called a digital system.

The basic memory element in logic circuits is the latch, which is constructed from two gates as shown in Figure 3.10(a), and which can be set or reset. A more useful variant is the D-type latch shown in (b) which remembers the state of the input at the time a separate clock either changes state for an edge-triggered device, or after it goes false for a level-triggered device. D-type latches are

Positive logic name	Boolean expression	Positive logic symbol	Positive logic truth table	Plain English
Inverter or NOT gate	$Q = \bar{A}$	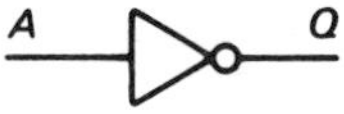	$\begin{array}{c\|c} A & Q \\ \hline 0 & 1 \\ 1 & 0 \end{array}$	Output is opposite of input
AND gate	$Q = A \cdot B$	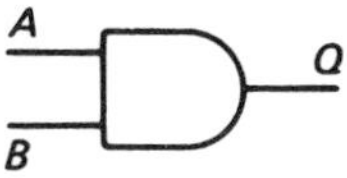	$\begin{array}{cc\|c} A & B & Q \\ \hline 0 & 0 & 0 \\ 0 & 1 & 0 \\ 1 & 0 & 0 \\ 1 & 1 & 1 \end{array}$	Output true when both inputs are true only
NAND (Not AND) gate	$Q = \overline{A \cdot B}$ $= \bar{A} + \bar{B}$	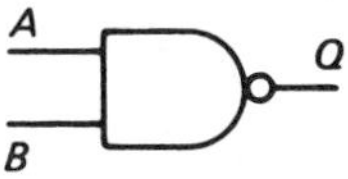	$\begin{array}{cc\|c} A & B & Q \\ \hline 0 & 0 & 1 \\ 0 & 1 & 1 \\ 1 & 0 & 1 \\ 1 & 1 & 0 \end{array}$	Output false when both inputs are true only
OR gate	$Q = A + B$	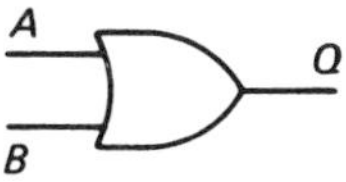	$\begin{array}{cc\|c} A & B & Q \\ \hline 0 & 0 & 0 \\ 0 & 1 & 1 \\ 1 & 0 & 1 \\ 1 & 1 & 1 \end{array}$	Output true if either or both inputs true
NOR (Not OR) gate	$Q = \overline{A + B}$ $= \bar{A} \cdot \bar{B}$	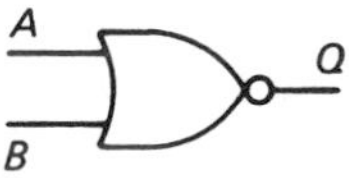	$\begin{array}{cc\|c} A & B & Q \\ \hline 0 & 0 & 1 \\ 0 & 1 & 0 \\ 1 & 0 & 0 \\ 1 & 1 & 0 \end{array}$	Output false if either or both inputs true
Exclusive OR (XOR) gate	$Q = A \oplus B$	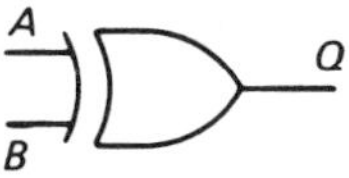	$\begin{array}{cc\|c} A & B & Q \\ \hline 0 & 0 & 0 \\ 0 & 1 & 1 \\ 1 & 0 & 1 \\ 1 & 1 & 0 \end{array}$	Output true if inputs are different

Figure 3.9 The basic logic gates compared.

commonly available with four or eight latches to the chip. A shift register can be made from a series of latches by connecting the Q output of one latch to the D input of the next and connecting all of the clock inputs in parallel. Data are delayed by the number of stages in the register. Shift registers are also useful for converting between serial and parallel data transmissions.

Where large numbers of bits are to be stored, cross-coupled latches are less suitable because they are more complicated to fabricate inside integrated circuits than dynamic memory, and consume more current.

In large random access memories (RAMs), the data bits are stored as the presence or absence of charge in a tiny capacitor as shown in Figure 3.10(c). The capacitor is formed by a metal electrode, insulated by a layer of silicon dioxide from a semiconductor substrate, hence the term MOS (metal oxide semiconductor). The charge will suffer leakage, and the value would become

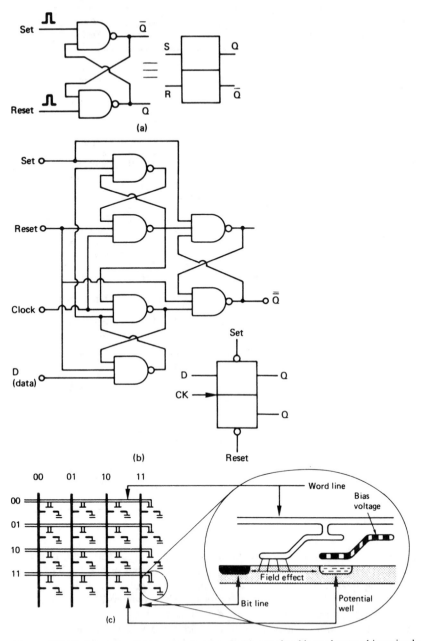

Figure 3.10 Digital semiconductor memory types. In (a), one data bit can be stored in a simple set–reset latch, which has little application because the D-type latch in (b) can store the state of the single data input when the clock occurs. These devices can be implemented with bipolar transistors or FETs, and are called static memories because they can store indefinitely. They consume a lot of power.

   In (c), a bit is stored as the charge in a potential well in the substrate of a chip. It is accessed by connecting the bit line with the field effect from the word line. The single well where the two lines cross can then be written or read. These devices are called dynamic RAMs because the charge decays, and they must be read and rewritten (refreshed) periodically.

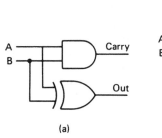

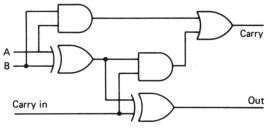

(a)

Data A	Bits B	Carry in	Out	Carry out
0	0	0	0	0
0	0	1	1	0
0	1	0	1	0
0	1	1	0	1
1	0	0	1	0
1	0	1	0	1
1	1	0	0	1
1	1	1	1	1

(b)

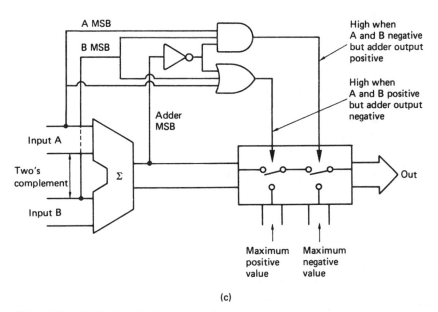

(c)

Figure 3.11 (a) Half adder; (b) full-adder circuit and truth table; (c) comparison of sign bits prevents wraparound on adder overflow by substituting clipping level.

indeterminate after a few milliseconds. Where the delay needed is less than this, decay is of no consequence, as data will be read out before they have had a chance to decay. Where longer delays are necessary, such memories must be refreshed periodically by reading the bit value and writing it back to the same place. Most modern MOS RAM chips have suitable circuitry built in. Large RAMs store millions of bits, and it is clearly impractical to have a connection to each one. Instead, the desired bit has to be addressed before it can be read or written. The size of the chip package restricts the number of pins available,

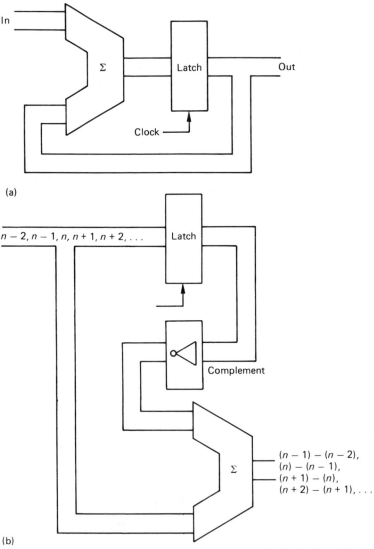

Figure 3.12 Two configurations which are common in processing. In (a) the feedback around the adder adds the previous sum to each input to perform accumulation or digital integration. In (b) an inverter allows the difference between successive inputs to be computed. This is differentiation.

so that large memories use the same address pins more than once. The bits are arranged internally as rows and columns, and the row address and the column address are specified sequentially on the same pins.

The circuitry necessary for adding pure binary or two's complement numbers is shown in Figure 3.11. Addition in binary requires 2 bits to be taken at a time from the same position in each word, starting at the least significant bit. Should both be ones, the output is zero, and there is a *carry-out* generated. Such a circuit is called a half adder, shown in Figure 3.11(a) and is suitable for the least significant bit of the calculation. All higher stages will require a circuit which can accept a carry input as well as two data inputs. This is known as a full adder (Figure 3.11(b)). Multibit full adders are available in chip form, and have carry-in and carry-out terminals to allow them to be cascaded to operate on long wordlengths. Such a device is also convenient for inverting a two's complement number, in conjunction with a set of inverters. The adder chip has one set of inputs grounded, and the carry-in permanently held true, such that it adds 1 to the one's complement number from the inverter.

When mixing by adding sample values, care has to be taken to ensure that if the sum of the two sample values exceeds the number range the result will be clipping rather than wraparound. In two's complement, the action necessary depends on the polarities of the two signals. Clearly if one positive and one negative number are added, the result cannot exceed the number range. If two positive numbers are added, the symptom of positive overflow is that the most significant bit sets, causing an erroneous negative result, whereas a negative overflow results in the most significant bit clearing. The overflow control circuit will be designed to detect these two conditions, and override the adder output. If the MSB of both inputs is zero, the numbers are both positive, thus if the sum has the MSB set, the output is replaced with the maximum positive code (0111 ...). If the MSB of both inputs is set, the numbers are both negative, and if the sum has no MSB set, the output is replaced with the maximum negative code (1000 ...). These conditions can also be connected to warning indicators. Figure 3.11(c) shows this system in hardware. The resultant clipping on overload is sudden, and sometimes a PROM is included which translates values around and beyond maximum to soft-clipped values below or equal to maximum.

A storage element can be combined with an adder to obtain a number of useful functional blocks which will crop up frequently in audio equipment. Figure 3.12(a) shows that a latch is connected in a feedback loop around an adder. The latch contents are added to the input each time it is clocked. The configuration is known as an accumulator in computation because it adds up or accumulates values fed into it. In filtering, it is known as a discrete-time integrator. If the input is held at some constant value, the output increases by that amount on each clock. The output is thus a sampled ramp.

Figure 3.12(b) shows that the addition of an invertor allows the difference between successive inputs to be obtained. This is digital differentiation. The output is proportional to the slope of the input.

3.4 Structure of a digital vision mixer

When making a digital recording, the gain of the analog input will usually be adjusted so that the quantizing range is fully exercised in order to make a recording of maximum signal-to-noise ratio. During post production, the recording may be played back and mixed with other signals. Effects such as dissolves and soft-edged wipes can only be achieved if the level of each input signal can be controlled independently. Gain is controlled in the digital domain by multiplying each sample value by a coefficient. If that coefficient is less than 1, attenuation will result; if it is greater than 1, amplification can be obtained.

Multiplication in binary circuits is difficult. It can be performed by repeated adding, but this is too slow to be of any use. In fast multiplication, one of the inputs will be simultaneously multiplied by 1, 2, 4, etc., by hard-wired bit shifting. Figure 3.13 shows that the other input bits will determine which of these powers will be added to produce the final sum, and which will be neglected. If multiplying by five, the process is the same as multiplying by four, multiplying by one, and adding the two products. This is achieved by adding the input to itself shifted two places. As the wordlength of such a device increases, the complexity increases exponentially, so this is a natural application for an integrated circuit. It is probably true that digital video would not have been viable without such chips.

In a digital mixer, the gain coefficients will originate in hand-operated faders, just as in analog. Analog faders may be retained and used to produce a

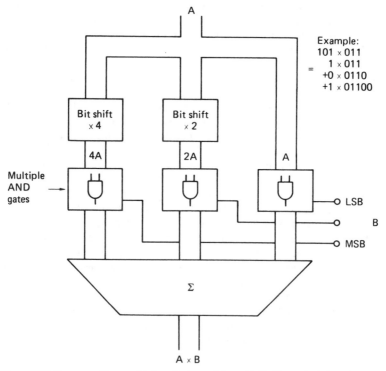

Figure 3.13 Structure of fast multiplier: the input A is multiplied by 1, 2, 4, 8, etc., by bit shifting. The digits of the B input then determine which multiples of A should be added together by enabling AND gates between the shifters and the adder. For long wordlengths, the number of gates required becomes enormous, and the device is best implemented in a chip.

varying voltage which is converted to a digital code or gain coefficient in an ADC, but it is also possible to obtain coefficients directly in digital faders. Digital faders are a form of displacement transducer in which the mechanical position of the control is converted directly to a digital code. The position of other controls, such as jog wheels on VTRs or editors, will also need to be digitized. Controls can be linear or rotary, and absolute or relative. In an absolute control, the position of the knob determines the output directly. In a relative control, the knob can be moved to increase or decrease the output, but its absolute position is meaningless.

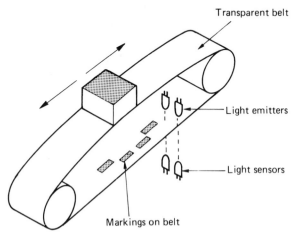

Figure 3.14 An absolute linear fader uses a number of light beams which are interrupted in various combinations according to the position of a grating. A Gray code shown in Figure 3.15 must be used to prevent false codes.

Figure 3.14 shows an absolute linear fader. A grating is moved with respect to several light beams, one for each bit of the coefficient required. The interruption of the beams by the grating determines which photocells are illuminated. It is not possible to use a pure binary pattern on the grating because this results in transient false codes due to mechanical tolerances. Figure 3.15 shows some examples of these false codes. For example, on moving the fader from 3 to 4, the MSB goes true slightly before the middle bit goes false. This results in a momentary value of $4 + 2 = 6$ between 3 and 4. The solution is to use a code in which only one bit ever changes in going from one value to the next. One such code is the Gray code which was devised to overcome timing hazards in relay logic but is now used extensively in position encoders.

Gray code can be converted to binary in a suitable PROM or gate array. These are available as industry-standard components.

Figure 3.16 shows a rotary incremental encoder. This produces a sequence of pulses whose number is proportional to the angle through which it has been turned. The rotor carries a radial grating over its entire perimeter. This turns over a second, fixed radial grating whose bars are not parallel to those of the first grating. The resultant moiré fringes travel inwards or outwards depending on the direction of rotation. Two suitably positioned light beams falling on photocells

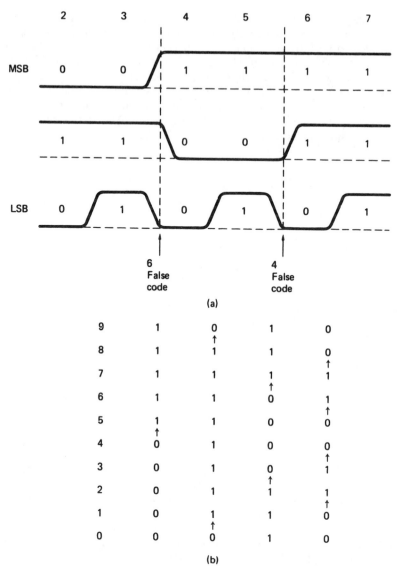

Figure 3.15 (a) Binary cannot be used for position encoders because mechanical tolerances cause false codes to be produced. (b) In Gray code, only one bit (arrowed) changes in between positions, so no false codes can be generated.

will produce outputs in quadrature. The relative phase determines the direction and the frequency is proportional to speed. The encoder outputs can be connected to a counter whose contents will increase or decrease according to the direction the rotor is turned. The counter provides the coefficient output.

The wordlength of the gain coefficients requires some thought as they determine the number of discrete gains available. If the coefficient wordlength is inadequate, the gain control becomes 'steppy' particularly towards the end of a

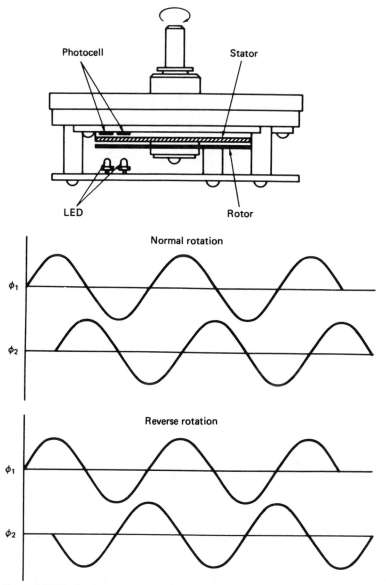

Figure 3.16 The fixed and rotating gratings produce moiré fringes which are detected by two light paths as quadrature sinusoids. The relative phase determines the direction, and the frequency is proportional to speed of rotation.

fadeout. A compromise between performance and the expense of high resolution faders is to insert a digital interpolator having a low-pass characteristic between the fader and the gain control stage. This will compute intermediate gains to higher resolution than the coarse fader scale so that the steps cannot be discerned.

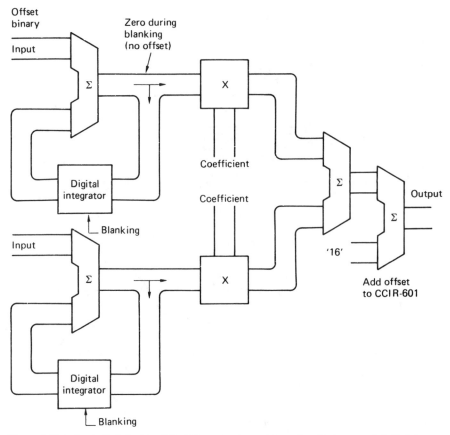

Figure 3.17 A simple digital mixer. Offset binary inputs must have the offset removed. A digital integrator will produce a counter-offset which is subtracted from every input sample. This will increase or reduce until the output of the subtractor is zero during blanking. The offset must be added back after processing if a CCIR-601 output is required.

The luminance path of a simple component digital mixer is shown in Figure 3.17. The CCIR-601 digital input is offset binary in that it has a nominal black level of 16_{10}, and a subtraction has to be made in order that fading will take place with respect to black. On a perfect signal, subtracting 16 would achieve this, but on a slightly out-of-range signal, it would not. Since the digital active line is slightly longer than the analog active line, the first sample should be blanking level, and this will be the value to subtract to obtain pure binary luminance with respect to black. This is the digital equivalent of black-level clamping. The two inputs are then multiplied by their respective coefficients, and added together to achieve the mix. Peak limiting will be required as in Section 3.3, and then, if the output is to be to CCIR-601, 16_{10} must be added to each sample value to establish the correct offset. In some video applications, a crossfade will be needed, and a rearrangement of the crossfading equation allows one multiplier to be used instead of two, as shown in Figure 3.18.

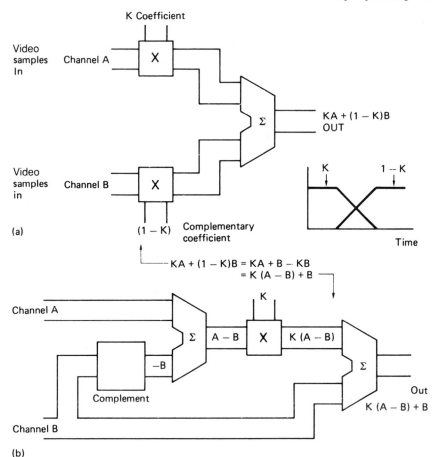

Figure 3.18 Crossfade (a) requires two multipliers. Reconfiguration (b) requires only one multiplier.

The colour difference signals are offset binary with an offset of 128_{10}, and again it is necessary to normalize these with respect to blanking level so that proper fading can be carried out. Since colour difference signals can be positive or negative, this process results in two's complement samples. Figure 3.19 shows some examples.

In this form, the samples can be added with respect to blanking level. Following addition, a limiting stage is used as before, and then, if it is desired to return to CCIR-601 standard, the MSB must be inverted once more in order to convert from two's complement to offset binary.

In practice the same multiplier can be used to process luminance and colour difference signals. Since these will be arriving time multiplexed at 27 MHz, it is only necessary to ensure that the correct coefficients are provided at the right time. Figure 3.20 shows an example of part of a slow fade. As the co-sited samples C_B, Y and C_R enter, all are multiplied by the same coefficient K_n, but the next sample will be luminance only, so this will be multiplied by K_{n+1}. The next

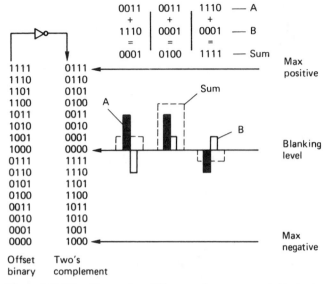

Figure 3.19 Offset binary colour difference values are converted to two's complement by reversing the state of the first bit. Two's complement values A and B will then add around blanking level.

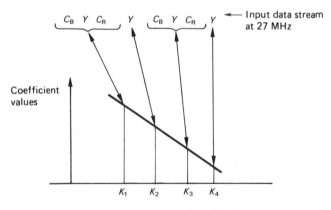

Figure 3.20 When using one multiplier to fade both luminance and colour difference in a 27 MHz multiplex 4:2:2 system, one coefficient will be used three times on the co-sited samples, whereas the next coefficient will only be used for a single luminance sample.

set of co-sited samples will be multiplied by K_{n+2} and so on. Clearly coefficients must be provided which change at 13.5 MHz. The sampling rate of the two inputs must be exactly the same, and in the same phase, or the circuit will not be able to add on a sample-by-sample basis. If the two inputs have come from different sources, they must be synchronized by the same master clock, and/or timebase correction must be provided on the inputs.

Some thought must be given to the wordlength of the system. If a sample is attenuated, it will develop bits which are below the radix point. For example, if an 8 bit sample is attenuated by 24 dB, the sample value will be shifted four places down. Extra bits must be available within the mixer to accommodate this shift. Digital vision mixers may have an internal wordlength of 16 bits or more. When several attenuated sources are added together to produce the final mix, the result will be a 16 bit sample stream. As the output will generally need to be of the same format as the input, the wordlength must be shortened. Shortening the wordlength of samples effectively makes the quantizing intervals larger and Chapter 2 showed that this can be called requantizing. This must be done using digital dithering to avoid artefacts.

It is possible to construct a digital composite switcher for NTSC or PAL which operates in much the same way as has been described for a colour difference channel above.

Figure 3.21 shows that the black level to be subtracted will be numerically larger than for the luminance signal in a component system, and may result in

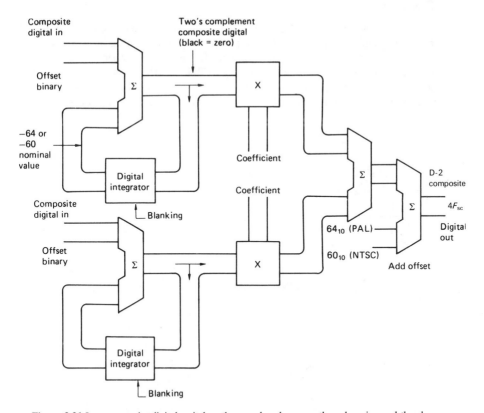

Figure 3.21 In a composite digital switcher, the samples also carry the subcarrier, and thus have meaning at levels below black. A larger offset, 60 or 64, is used than for components, and when this is subtracted a two's complement system is necessary to convey signals above or below black as positive or negative values which can then be faded without offsets resulting.

negative sample values which will need to be expressed in two's complement form before processing.

Clearly the adder which mixes the two channels can only produce a meaningful result if both input values are derived from samples which were taken at precisely the same phase relationship to subcarrier. The critical timing-in process of an analog composite switcher has been removed as far as the digital switcher is concerned, but the need for critical timing accuracy has not been eliminated; it has simply been moved to a different part of the system. A digital timebase corrector can be used to remove timing errors between input channels, but as it can only correct to the nearest sample, which is 90° in a $4F_{sc}$ system, it cannot remove errors due to the sampling clocks of the two channels having a different relationship to subcarrier. If the two channels have such a timing error, it can only be corrected in the digital domain by an interpolator of some complexity. It is necessary to bear in mind that a composite digital recorder must have an input sampling clock which is accurately set to a standardized phase relationship with subcarrier, otherwise the recordings made cannot be mixed with others in the digital domain.

3.5 Keying

Keying is the process where one video signal can be cut into another to replace part of the picture with a different image. One application of keying is where a switcher can wipe from one input to another using one of a variety of different patterns. Figure 3.22 shows that an analog switcher performs such an effect by generating a binary switching waveform in a pattern generator. Video switching between inputs actually takes place during the active line. In most analog

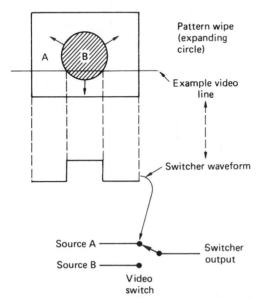

Figure 3.22 In a video switcher a pattern generator produces a switching waveform which changes from line to line and from frame to frame to allow moving pattern wipes between sources.

switchers, the switching waveform is digitally generated, then fed to a DAC whereas in a digital switcher, the pattern generator outputs become the coefficients supplied to the crossfader, which is sometimes referred to as a cutter. The switching edge must be positioned to an accuracy of a few nanoseconds, much less than the spacing of the pixels, otherwise slow wipes will not appear to move smoothly, and diagonal wipes will have stepped edges, a phenomenon known as ratcheting.

Positioning the switch point to sub-pixel accuracy is not particularly difficult, as Figure 3.23 shows. A suitable series of coefficients can position the effective crossover point anywhere. The finite slope of the coefficients results in a brief

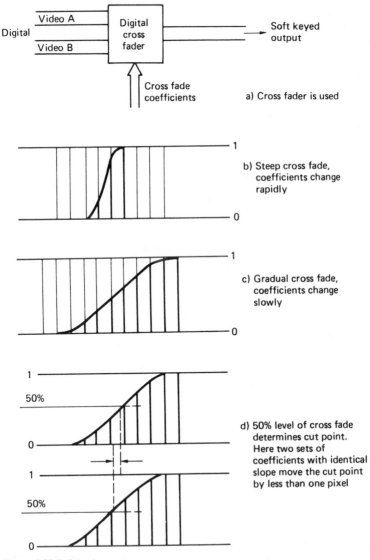

Figure 3.23 Soft keying.

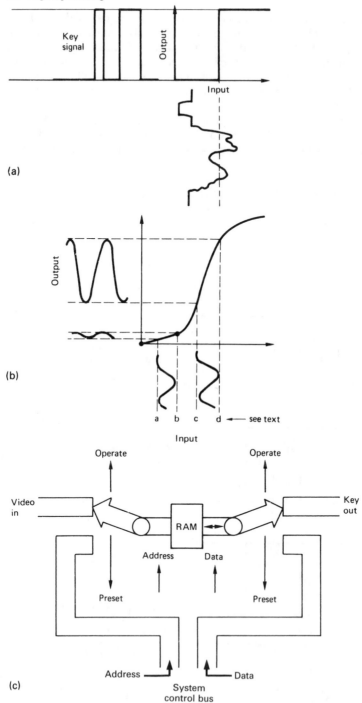

(a)

(b)

(c)

Figure 3.24 (a) A non-linear transfer function can be used to produce a keying signal. (b) The non-linear transfer function emphasizes contrast in part of the range but reduces it at other parts. (c) If a RAM is used as a flexible transfer function, it will be necessary to provide multiplexers so that the RAM can be preset with the desired values from the control system.

crossfade from one video signal to the other. This soft keying gives a much more realistic effect than binary switchers, which often give a 'cut out with scissors' appearance. In some machines the slope of the crossfade can be adjusted to achieve the desired degree of softness.

Another application of keying is to derive the switching signal by processing video from a camera in some way. By analysing colour difference signals, it is possible to determine where in a picture a particular colour occurs. When a key signal is generated in this way, the process is known as chroma keying, which is the electronic equivalent of matting in film.

In a 4:2:2 component system, It will be necessary to provide coefficients to the luminance crossfader at 13.5 MHz. Chroma samples only occur at half this frequency, so it is necessary to provide a chroma interpolator to raise the chroma sampling rate artificially. For chroma keying a simple linear interpolator is perfectly adequate. Intermediate chroma samples are simply the average of two adjacent samples.

As with analog switchers, chroma keying is also possible with composite digital inputs, but decoding must take place before it is possible to obtain the key signals. The video signals which are being keyed will, however, remain in the composite digital format.

In switcher/keyers, it is necessary to obtain a switching signal which ramps between two states from an input signal which can be any allowable video waveform. Manual controls are provided so that the operator can set thresholds and gains to obtain the desired effect. In the analog domain, these controls distort the transfer function of a video amplifier so that it is no longer linear. A digital keyer will perform the same functions using logic circuits.

Figure 3.24(a) shows that the effect of a non-linear transfer function is to switch when the input signal passes through a particular level. The transfer function is implemented in a memory in digital systems. The incoming video sample value acts as the memory address, so that the selected memory location is proportional to the video level. At each memory location, the appropriate output level code is stored. If, for example, each memory location stored its own address, the output would equal the input, and the device would be transparent. In practice switching is obtained by distorting the transfer function to obtain more gain in one particular range of input levels at the expense of less gain at other input levels. With the transfer function shown in Figure 3.24(b), an input level change from a to b causes a smaller output change, whereas the same level change between c and d causes a considerable output change.

If the memory is RAM, different transfer functions can be loaded in by the control system, and this requires multiplexers in both data and address lines as shown in Figure 3.24(c).

In practice such a RAM will be installed in Y, C_r and C_b channels, and the results will be combined to obtain the final switching coefficients.

3.6 Filters

Filtering is inseparable from digital video and audio. Analog or digital filters, and sometimes both, are required in ADCs, DACs, in the data channels of digital recorders and transmission systems and in sampling rate converters and equalizers. Optical systems used in disk recorders also act as filters.[1] There are many parallels between analog, digital and optical filters, which this section treats as a common subject. The main difference between analog and digital filters is that in the digital domain very complex architectures can be constructed at low cost in LSI and that arithmetic calculations are not subject to component tolerance or drift.

Filtering may modify the frequency response of a system, and/or the phase response. Every combination of frequency and phase response determines the impulse response in the time domain. Figure 3.25 shows that impulse response testing tells a great deal about a filter. In a perfect filter, all frequencies should experience the same time delay. If some groups of frequencies experience a different delay to others, there is a group-delay error. As an impulse contains an infinite spectrum, a filter suffering from group-delay error will separate the different frequencies of an impulse along the time axis.

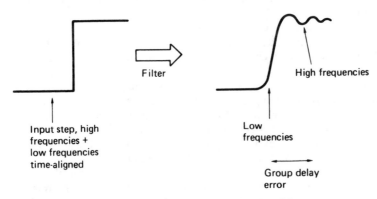

Figure 3.25 Group delay time-displaces signals as a function of frequency.

A pure delay will cause a phase shift proportional to frequency, and a filter with this characteristic is said to be phase linear. The impulse response of a phase-linear filter is symmetrical. If a filter suffers from group-delay error it cannot be phase linear. It is almost impossible to make a perfectly phase-linear analog filter, and many filters have a group-delay equalization stage following them which is often as complex as the filter itself. In the digital domain it is straightforward to make a phase-linear filter, and phase equalization becomes unnecessary.

Because of the sampled nature of the signal, whatever the response at low frequencies may be, all digital channels (and sampled analog channels) act as low-pass filters cutting off at the Nyquist limit, or half the sampling frequency.

Figure 3.26(a) shows a simple *RC* network and its impulse response. This is the familiar exponential decay due to the capacitor discharging through the resistor (in series with the source impedance which is assumed here to be negligible). The figure also shows the response to a squarewave in (b). These responses can be calculated because the inputs involved are relatively simple. When the input waveform and the impulse response are complex functions, this approach becomes almost impossible.

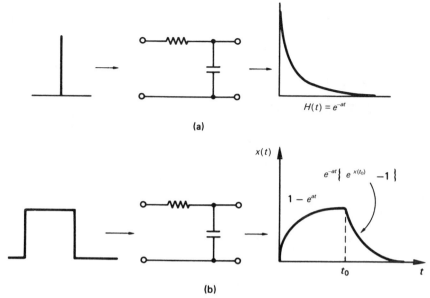

$$H(t) = e^{-at}$$

(a)

$x(t)$

$e^{-at}\{\ e^{\ x(t_0)}\quad -1\ \}$

$1 - e^{at}$

t_0

t

(b)

Figure 3.26 (a) The impulse response of a simple *RC* network is an exponential decay. This can be used to calculate the response to a square wave, as in (b).

In any filter, the time-domain output waveform represents the convolution of the impulse response with the input waveform. Convolution can be followed by reference to a graphic example in Figure 3.27. Where the impulse response is asymmetrical, the decaying tail occurs *after* the input. As a result it is necessary to reverse the impulse response in time so that it is mirrored prior to sweeping it through the input waveform. The output voltage is proportional to the shaded area shown where the two impulses overlap.

The same process can be performed in the sampled, or discrete, time domain as shown in Figure 3.28. The impulse and the input are now a set of discrete samples which clearly must have the same sample spacing. The impulse response only has value where impulses coincide. Elsewhere it is zero. The impulse response is therefore stepped through the input one sample period at a time. At each step, the area is still proportional to the output, but as the time steps are of uniform width, the area is proportional to the impulse height and so the output is obtained by adding up the lengths of overlap. In mathematical terms, the output samples represent the convolution of the input and the impulse response by summing the coincident cross products.

As a digital filter works in this way, perhaps it is not a filter at all, but just a mathematical simulation of an analog filter. This approach is quite useful in visualizing what a digital filter does.

Filters can be described in two main classes, as shown in Figure 3.29, according to the nature of the impulse response. Finite-impulse response (FIR) filters are always stable and, as their name suggests, respond to an impulse once, as they have only a forward path. In the temporal domain, the time for which the filter responds to an input is finite, fixed and readily established. The

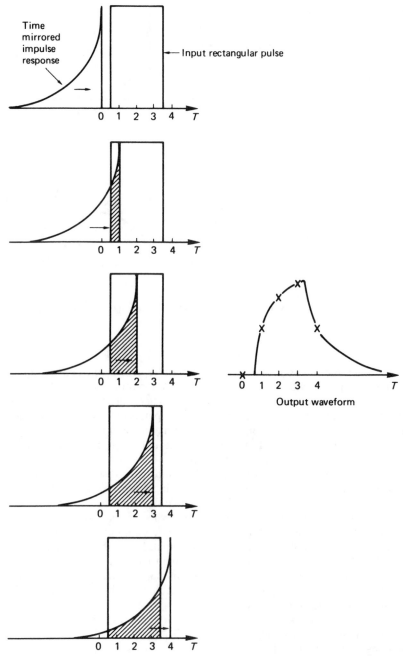

Figure 3.27 In the convolution of two continuous signals (the impulse response with the input), the impulse must be time reversed or mirrored. This is necessary because the impulse will be moved from left to right, and mirroring gives the impulse the correct time-domain response when it is moved past a fixed point. As the impulse response slides continuously through the input waveform, the area where the two overlap determines the instantaneous output amplitude. This is shown for five different times by the crosses on the output waveform.

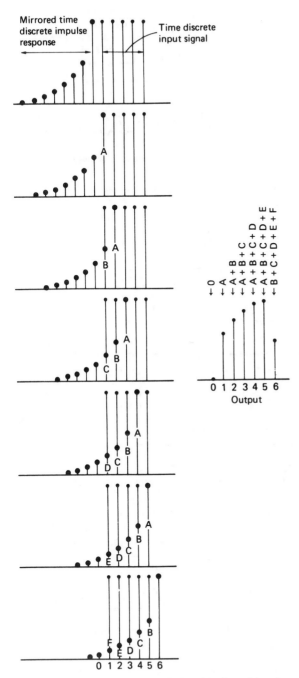

Figure 3.28 In time discrete convolution, the mirrored impulse response is stepped through the input one sample period at a time. At each step, the sum of the cross-products is used to form an output value. As the input in this example is a constant height pulse, the output is simply proportional to the sum of the coincident impulse response samples. This figure should be compared with Figure 3.27.

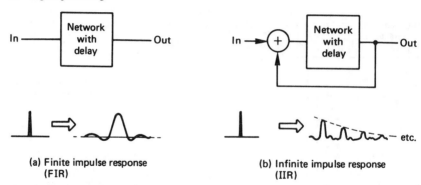

(a) Finite impulse response
(FIR)

(b) Infinite impulse response
(IIR)

Figure 3.29 An FIR filter (a) responds only once to an input, whereas the output of an IIR filter (b) continues indefinitely rather like a decaying echo.

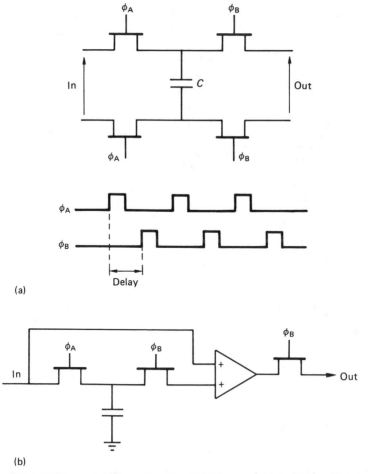

(a)

(b)

Figure 3.30 In a switched capacitor delay (a), there are two clock phases, and during the first the input voltage is transferred to the capacitor. During the second phase the capacitor voltage is transferred to the output. (b) A simple switched capacitor filter. The delay causes a phase shift which is dependent on frequency and the resultant frequency response is sinusoidal.

same is therefore true about the distance over which an FIR filter responds in the spatial domain. FIR filters can be made perfectly phase linear if required. Most filters used for sampling-rate conversion and oversampling fall into this category.

Infinite-impulse response (IIR) filters respond to an impulse indefinitely and are not necessarily stable, as they have a return path from the output to the input. For this reason they are also called recursive filters. As the impulse response is not symmetrical, IIR filters are not phase linear and are little used in video.

Somewhere between the analog filter and the digital filter is the switched capacitor filter. This uses analog quantities, namely the charges on capacitors, but the time axis is discrete because the various charges are routed using electronic switches which close during various phases of the sampling-rate clock. Switched capacitor filters have the same characteristics as digital filters with infinite precision. They are often used in preference to continuous-time analog filters in integrated circuit converters because they can be implemented with the same integration techniques. Figure 3.30(a) shows a switched capacitor delay. There are two clock phases and during the first the input voltage is transferred to the capacitor. During the second phase the capacitor voltage is transferred to the output. Combining delay with operational amplifier summation allows frequency-dependent circuitry to be realized. Figure 3.30(b) shows a simple switched capacitor filter. The delay causes a phase shift which is dependent on frequency. The frequency response is sinusoidal.

3.7 Transforms

Convolution is a lengthy process to perform on paper. It is much easier to work in the frequency domain. Figure 3.31 shows that if a signal with a spectrum or frequency content a is passed through a filter with a frequency response b the result will be an output spectrum which is simply the product of the two. If the frequency responses are drawn on logarithmic scales (i.e. calibrated in dB) the two can be simply added because the addition of logs is the same as multiplication. Whilst frequency in audio has traditionally meant temporal frequency measured in Hertz, frequency in video can also be spatial and measured in lines per millimetre (mm^{-1}). Multiplying the spectra of the responses is a much simpler process than convolution.

In order to move to the frequency domain or spectrum from the time domain or waveform, it is necessary to use the Fourier transform, or in sampled systems, the discrete Fourier Transform (DFT). Fourier analysis holds that any waveform can be reproduced by adding together an arbitrary number of harmonically related sinusoids of various amplitudes and phases. The spectrum can be drawn by plotting the amplitude of the harmonics against frequency. It will be seen that this gives a spectrum which is a decaying wave. It passes through zero at all even multiples of the fundamental. The shape of the spectrum is a $\sin x/x$ curve. If a square wave has a $\sin x/x$ spectrum, it follows that a filter with a rectangular impulse response will have a $\sin x/x$ spectrum.

A low-pass filter has a rectangular spectrum, and this has a $\sin x/x$ impulse response. These characteristics are known as a transform pair. In transform pairs, if one domain has one shape of the pair, the other domain will have the other

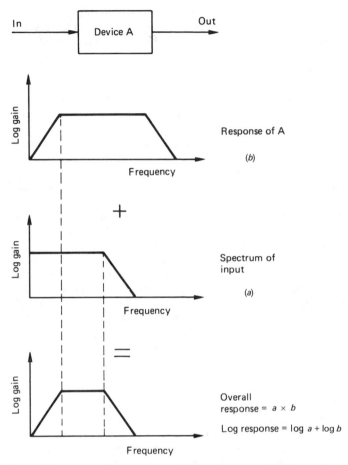

Figure 3.31 In the frequency domain, the response of two series devices is the product of their individual responses at each frequency. On a logarithmic scale the responses are simply added.

shape. Thus a square wave has a $\sin x/x$ spectrum and a $\sin x/x$ impulse has a square spectrum. Figure 3.32 shows a number of transform pairs. Note the pulse pair. A time-domain pulse of infinitely short duration has a flat spectrum. Thus a flat waveform, i.e. DC, has only zero in its spectrum. Interestingly the transform of a Gaussian response is still Gaussian. The impulse response of the optics of a laser disk has a $\sin x^2/x^2$ function, and this is responsible for the triangular falling frequency response of the pickup.

The Fourier transform is a processing technique which analyses signals changing with respect to time or distance and expresses them in the form of a temporal or spatial spectrum. Any waveform can be broken down into frequency components. Figure 3.33 shows that if the amplitude and phase of each frequency component is known, linearly adding the resultant components in an inverse transform results in the original waveform.

The Fourier transform may be performed on a continuous analog waveform, in which case a continuous spectrum results. However, in digital systems the

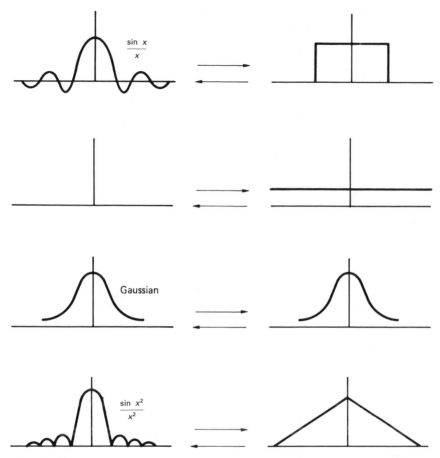

Figure 3.32 The concept of transform pairs illustrates the duality of the frequency (including spatial frequency) and time domains.

waveform is expressed as a number of discrete samples. As a result the Fourier transform analyses the signal into an equal number of discrete frequencies. This is known as a discrete Fourier transform or DFT. The fast Fourier transform is no more than an efficient way of computing the DFT.[2]

It will be evident from Figure 3.33 that knowledge of the phase of the frequency component is vital, as changing the phase of any component will seriously alter the reconstructed waveform. Thus the DFT must accurately analyse the phase of the signal components.

There are a number of ways of expressing phase. Figure 3.34 shows a point which is rotating about a fixed axis at constant speed. Looked at from the side, the point oscillates up and down at constant frequency. The waveform of that motion is a sine wave, and that is what we would see if the rotating point were to translate along its axis whilst we continued to look from the side.

One way of defining the phase of a waveform is to specify the angle through which the point has rotated at time zero ($T = 0$). If a second point is made to

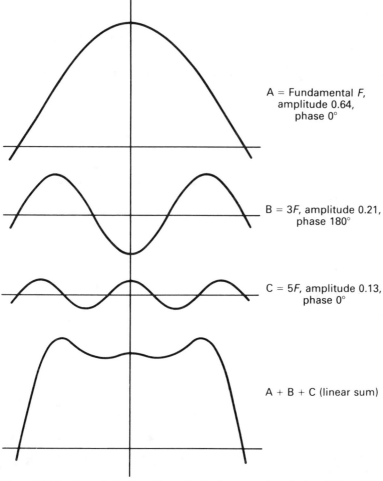

A = Fundamental *F*,
amplitude 0.64,
phase 0°

B = 3*F*, amplitude 0.21,
phase 180°

C = 5*F*, amplitude 0.13,
phase 0°

A + B + C (linear sum)

Figure 3.33 Fourier analysis allows the synthesis of any waveform by the addition of discrete frequencies of appropriate amplitude and phase.

revolve at 90° to the first, it would produce a cosine wave when translated. It is possible to produce a waveform having arbitrary phase by adding together the sine and cosine waves in various proportions and polarities. For example adding the sine and cosine waves in equal proportion results in a waveform lagging the sine wave by 45 degrees.

Figure 3.34 shows that the proportions necessary are respectively the sine and the cosine of the phase angle. Thus the two methods of describing phase can be readily interchanged.

The discrete Fourier transform spectrum-analyses a block of samples by searching separately for each discrete target frequency. It does this by multiplying the input waveform by a sine wave having the target frequency and adding up or integrating the products. Figure 3.35(a) shows that multiplying by the target frequency gives a non-zero integral when the input frequency is the

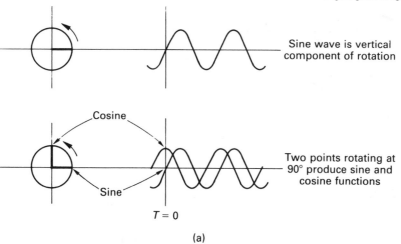

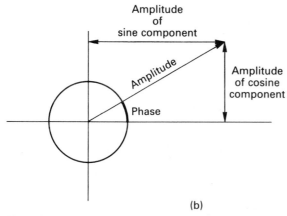

(a)

(b)

Figure 3.34 The origin of sine and cosine waves is to take a particular viewpoint of a rotation. Any phase can be synthesized by adding proportions of sine and cosine waves.

same, whereas Figure 3.35(b) shows that with a different input frequency (in fact all other different frequencies) the integral is zero showing that no component of the target frequency exists. Thus from a real waveform containing many frequencies all frequencies except the target frequency are excluded. The magnitude of the integral is proportional to the amplitude of the target component.

Figure 3.35(c) shows that the target frequency will not be detected if it is phase shifted 90 degrees as the product of quadrature waveforms is always zero. Thus the discrete Fourier transform must make a further search for the target frequency using a cosine wave. It follows from the arguments above that the relative proportions of the sine and cosine integrals reveal the phase of the input component. Thus each discrete frequency in the spectrum must be the result of a pair of quadrature searches.

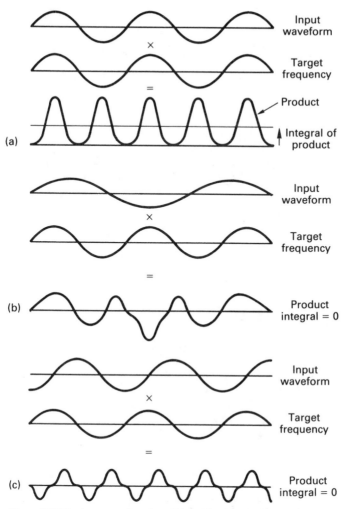

Figure 3.35 The input waveform is multiplied by the target frequency and the result is averaged or integrated. In (a) the target frequency is present and a large integral results. With another input frequency the integral is zero as in (b). The correct frequency will also result in a zero integral shown in (c) if it is at 90° to the phase of the search frequency. This is overcome by making two searches in quadrature.

Searching for one frequency at a time as above will result in a DFT, but only after considerable computation. However, a lot of the calculations are repeated many times over in different searches. The fast Fourier transform gives the same result with less computation by logically gathering together all of the places where the same calculation is needed and making the calculation once.

The discrete cosine transform (DCT) is a special case of a discrete Fourier transform in which the sine components of the coefficients have been eliminated leaving a single number. This is actually quite easy. Figure 3.36(a) shows the input samples to a transform process. By repeating the samples in a time-reversed order and performing a discrete Fourier transform on the double-length sample

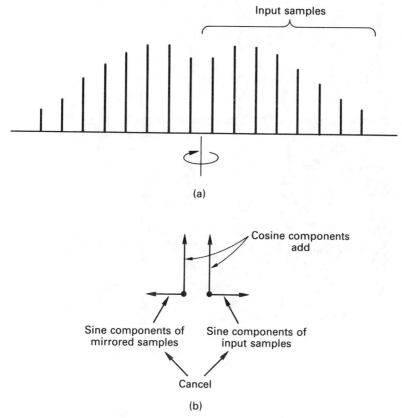

Figure 3.36 The DCT is obtained by mirroring the input block as shown in (a) prior to an FFT. The mirroring cancels out the sine components as in (b), leaving only cosine coefficients.

set a DCT is obtained. The effect of mirroring the input waveform is to turn it into an even function whose sine coefficients are all zero. The result can be understood by considering the effect of individually transforming the input block and the reversed block. Figure 3.36(b) shows that the phases of all the components of one block are in the opposite sense to those in the other. This means that when the components are added to give the transform of the double-length block all of the sine components cancel out, leaving only the cosine coefficients; hence the name of the transform.[3] In practice the sine component calculation is eliminated. Another advantage is that doubling the block length by mirroring doubles the frequency resolution, so that twice as many useful coefficients are produced. In fact a DCT produces as many useful coefficients as input samples.

For image processing two-dimensional transforms are needed. In this case for every horizontal frequency, a search is made for all possible vertical frequencies. A two-dimensional DCT is shown in Figure 3.37. The DCT is separable, which means that the two-dimensional DCT can be obtained by computing in each dimension separately. Fast DCT algorithms are available.[4]

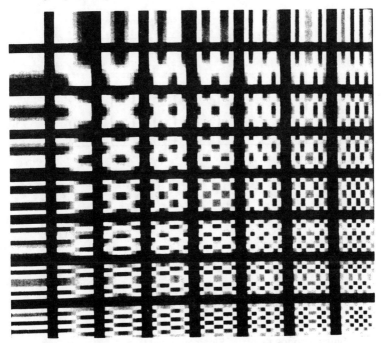

Figure 3.37 The discrete cosine transform breaks up an image area into discrete frequencies in two dimensions. The lowest frequency can be seen here at the top left corner. Horizontal frequency increases to the right and vertical frequency increases downwards.

The DCT is primarily used in data reduction processing. The DCT itself does not result in any reduction, as there are as many coefficients as samples, but it converts the input video into a form where redundancy can be easily detected and removed.

3.8 FIR filters

An FIR filter works by graphically constructing the impulse response for every input sample. It is first necessary to establish the correct impulse response. Figure 3.38(a) shows an example of a low-pass filter which cuts off at one-quarter of the sampling rate. The impulse response of a perfect low-pass filter is a sin x/x curve, where the time between the two central zero crossings is the reciprocal of the cut-off frequency. According to the mathematics, the waveform has always existed, and carries on for ever. The peak value of the output coincides with the input impulse. This means that the filter is not causal, because the output has changed before the input is known. Thus in all practical applications it is necessary to truncate the extreme ends of the impulse response, which causes an aperture effect, and to introduce a time delay in the filter equal to half the duration of the truncated impulse in order to make the filter causal. As an input impulse is shifted through the series of registers in Figure 3.38(b), the impulse response is created, because at each point it is multiplied by a coefficient as in Figure 3.38(c). These coefficients are simply the result of sampling and quantizing the desired impulse

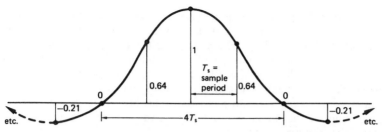

Figure 3.38(a) The impulse response of an LPF is a sin x/x curve which stretches from −∞ to +∞ in time. The ends of the response must be neglected, and a delay introduced to make the filter causal.

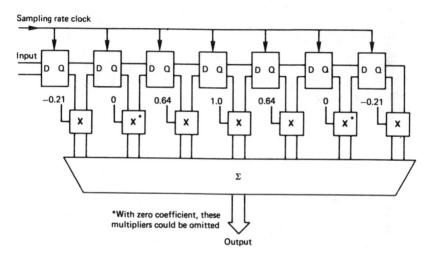

Figure 3.38(b) The structure of an FIR LPF. Input samples shift across the register and at each point are multiplied by different coefficients.

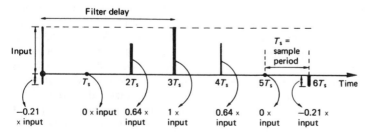

Figure 3.38(c) When a single unit sample shifts across the circuit of Figure 3.38(b), the impulse response is created at the output as the impulse is multiplied by each coefficient in turn.

response. Clearly the sampling rate used to sample the impulse must be the same as the sampling rate for which the filter is being designed. In practice the coefficients are calculated, rather than attempting to sample an actual impulse response. The coefficient wordlength will be a compromise between cost and performance. Because the input sample shifts across the system registers to create the shape of the impulse response, the configuration is also known as a transversal filter. In operation with real sample streams, there will be several consecutive sample values in the filter registers at any time in order to convolve the input with the impulse response.

Simply truncating the impulse response causes an abrupt transition from input samples which matter to those which do not. Truncating the filter superimposes a rectangular shape on the time-domain impulse response. In the frequency domain the rectangular shape transforms to a $\sin x/x$ characteristic which is superimposed on the desired frequency response as a ripple. One consequence of this is known as Gibb's phenomenon: a tendency for the response to peak just before the cut-off frequency.[5,6] As a result, the length of the impulse which must be considered will depend not only on the frequency response, but also on the amount of ripple which can be tolerated. If the relevant period of the impulse is measured in sample periods, the result will be the number of points or multiplications needed in the filter. Figure 3.39 compares the performance of filters with different numbers of points. Video filters may use as few as eight points whereas a high-quality digital audio FIR filter may need as many as 96 points.

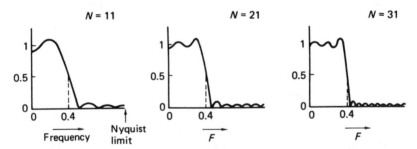

Figure 3.39 The truncation of the impulse in an FIR filter caused by the use of a finite number of points (*N*) results in ripple in the response. Shown here are three different numbers of points for the same impulse response. The filter is an LPF which rolls off at 0.4 of the fundamental interval. (Courtesy *Philips Technical Review*).

Rather than simply truncating the impulse response in time, it is better to make a smooth transition from samples which do not count to those that do. This can be done by multiplying the coefficients in the filter by a window function which peaks in the centre of the impulse. In the example of Figure 3.40, the low-pass filter of Figure 3.38 is shown with a Bartlett window. Acceptable ripple determines the number of significant sample periods embraced by the impulse. This determines in turn both the number of points in the filter and the filter delay. As the impulse is symmetrical, the delay will be half the impulse period. The impulse response is a $\sin x/x$ function, and this has been calculated in the figure. The $\sin x/x$ response is next multiplied by the window function to give the windowed impulse response.

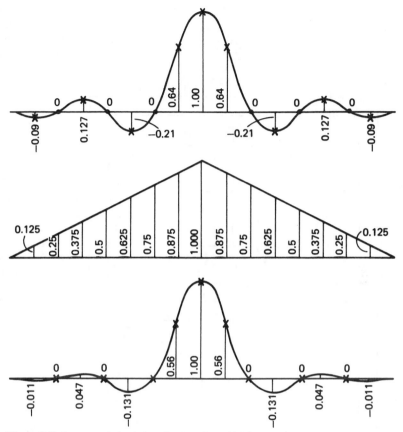

Figure 3.40 A truncated sin x/x impulse (top) is multiplied by a Bartlett window function (centre) to produce the actual coefficients used (bottom).

If the coefficients are not quantized finely enough, it will be as if they had been calculated inaccurately, and the performance of the filter will be less than expected. Figure 3.41 shows an example of quantizing coefficients. Conversely, raising the wordlength of the coefficients increases cost.

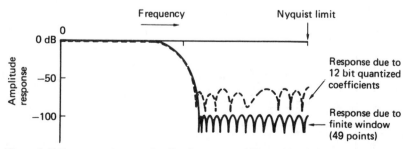

Figure 3.41 Frequency response of a 49 point transversal filter with infinite precision (solid line) shows ripple due to finite window size. Quantizing coefficients to 12 bits reduces attenuation in the stopband. (Responses courtesy *Philips Technical Review*).

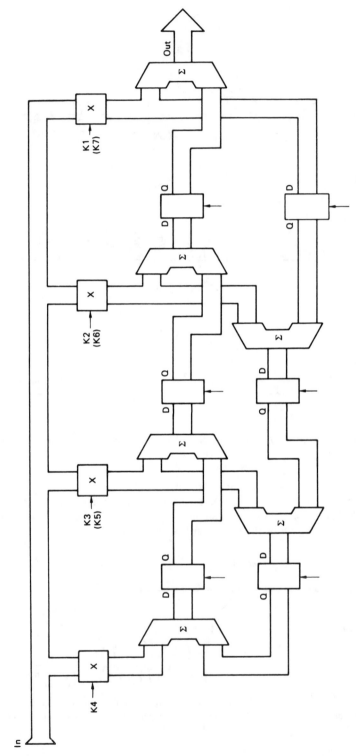

Figure 3.42 A seven-point folded filter for a symmetrical impulse response. In this case K1 and K7 will be identical, and so the input sample can be multiplied once, and the product fed into the output shift system in two different places. The centre coefficient K4 appears once. In an even-numbered filter the centre coefficient would also be used twice.

The FIR structure is inherently phase linear because it is easy to make the impulse response absolutely symmetrical. The individual samples in a digital system do not know in isolation what frequency they represent, and they can only pass through the filter at a rate determined by the clock. Because of this inherent phase linearity, an FIR filter can be designed for a specific impulse response, and the frequency response will follow.

The frequency response of the filter can be changed at will by changing the coefficients. A programmable filter only requires a series of PROMs to supply the coefficients; the address supplied to the PROMs will select the response. The frequency response of a digital filter will also change if the clock rate is changed, so it is often less ambiguous to specify a frequency of interest in a digital filter in terms of a fraction of the fundamental interval rather than in absolute terms. The configuration shown in Figure 3.38 serves to illustrate the principle. The units used on the diagrams are sample periods and the response is proportional to these periods or spacings, and so it is not necessary to use actual figures.

Where the impulse response is symmetrical, it is often possible to reduce the number of multiplications, because the same product can be used twice, at equal distances before and after the centre of the window. This is known as folding the filter. A folded filter is shown in Figure 3.42.

FIR filters can be used for interpolation, which in its most general sense is the computation of a sample value which lies somewhere between existing samples. Interpolation is an important enabling technology on which a large number of practical digital video devices are based. There are three basic but related categories of interpolation, as shown in Figure 3.43. The most straightforward (a) changes the sample spacing by an integer ratio, up or down. The timing of the system is thus simplified because all samples (input and output) are present on edges of the higher-rate sampling clock. Such a system is generally adopted for oversampling converters; the exact sampling rate immediately adjacent to the analog domain is not critical, and will be chosen to make the filters easier to implement.

Next in order of difficulty is the category shown in (b) where the rate is changed by the ratio of two small integers. Samples in the input periodically time-align with the output. Such devices can be used for converting from $4 \times F_{sc}$ to $3 \times F_{sc}$, in the vertical processing of standards converters, or between the various rates of CCIR-601.

The most complex interpolation category is where there is no simple relationship between input and output sampling rates, and in fact they may vary. This situation, shown in (c), is known as variable-ratio conversion. The temporal or spatial relationship of input and output samples is arbitrary. This problem will be met in effects machines which zoom or rotate images and in motion-compensated standards converters.

In considering how interpolators work it should be recalled that, according to sampling theory, all sampled systems have finite bandwidth. An individual digital sample value is obtained by sampling the instantaneous voltage of the original analog waveform, and because it has zero duration, it must contain an infinite spectrum. However, such a sample can never be seen or heard in that form because of the reconstruction process, which limits the spectrum of the impulse to the Nyquist limit. After reconstruction, one infinitely short digital sample ideally represents a $\sin x/x$ pulse whose central peak width is determined by the response of the reconstruction filter, and whose amplitude is proportional to the

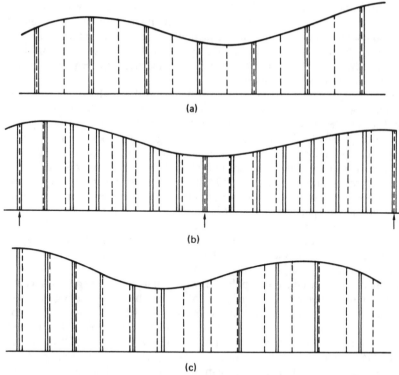

Figure 3.43 Categories of rate conversion. (a) Integer-ratio conversion, where the lower-rate samples are always coincident with those of the higher rate. There are a small number of phases needed. (b) Fractional-ratio conversion, where sample coincidence is periodic. A larger number of phases is required. Example here is conversion from 50.4 kHz to 44.1 kHz (8/7). (c) Variable-ratio conversion, where there is no fixed relationship, and a large number of phases are required.

sample value. This implies that, in reality, one sample value has meaning over á considerable timespan, rather than just at the sample instant. If this were not true, it would be impossible to build an interpolator.

As in rate reduction, performing the steps separately is inefficient. The bandwidth of the information is unchanged when the sampling rate is increased; therefore the original input samples will pass through the filter unchanged, and it is superfluous to compute them. The combination of the two processes into an interpolating filter minimizes the amount of computation.

As the purpose of the system is purely to change the sample spacing, the filter must be as transparent as possible, and this implies that a linear-phase configuration is mandatory, suggesting the use of an FIR structure. Figure 3.44 shows that the theoretical impulse response of such a filter is a $\sin x/x$ curve which has zero value at the position of adjacent input samples. In practice this impulse cannot be implemented because it is infinite. The impulse response used will be truncated and windowed as described earlier. To simplify this discussion, assume that a $\sin x/x$ impulse is to be used. There is a strong parallel with the

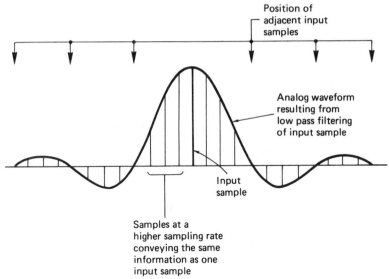

Figure 3.44 A single sample results in a sin x/x waveform after filtering in the analog domain. At a new, higher, sampling rate, the same waveform after filtering will be obtained if the numerous samples of differing size shown here are used. It follows that the values of these new samples can be calculated from the input samples in the digital domain in an FIR filter.

operation of a DAC where the analog voltage is returned to the time-continuous state by summing the analog impulses due to each sample. In a digital interpolating filter, this process is duplicated.[7]

If, for example, the sampling rate is to be doubled, new samples must be interpolated exactly half-way between existing samples. The necessary impulse response is shown in Figure 3.45; it can be sampled at the *output* sample period and quantized to form coefficients. If a single input sample is multiplied by each of these coefficients in turn, the impulse response of that sample at the new sampling rate will be obtained. Note that every other coefficient is zero, which confirms that no computation is necessary on the existing samples; they are just transferred to the output. The intermediate sample is computed by adding together the impulse responses of every input sample in the window. The figure shows how this mechanism operates. If the sampling rate is to be increased by a factor of four, three sample values must be interpolated between existing input samples. Figure 3.46 shows that it is only necessary to sample the impulse response at one-quarter the period of input samples to obtain three sets of coefficients which will be used in turn. In hardware-implemented filters, the input sample which is passed straight to the output is transferred by using a fourth filter phase where all coefficients are zero except the central one which is unity.

In a variable-ratio interpolator, values will exist for the points at which input samples were made, but it is necessary to compute what the sample values would have been at absolutely any point between available samples. The general concept of the interpolator is the same as for the fractional-ratio converter, except

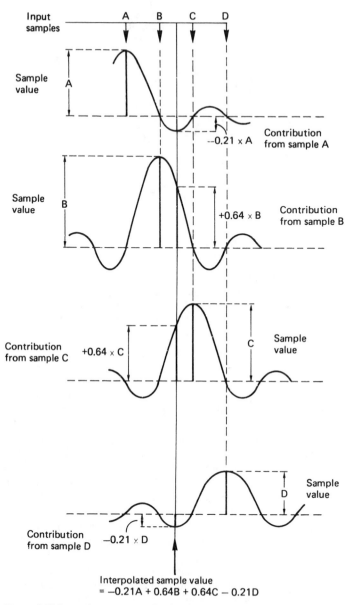

Figure 3.45 A two times oversampling interpolator. To compute an intermediate sample, the input samples are imagined to be sin x/x impulses, and the contributions from each at the point of interest can be calculated. In practice, rather more samples on either side need to be taken into account.

that an infinite number of filter phases is ideally necessary. Since a realizable filter will have a finite number of phases, it is necessary to study the degradation this causes. The desired continuous temporal or spatial axis of the interpolator is quantized by the phase spacing, and a sample value needed at a particular point

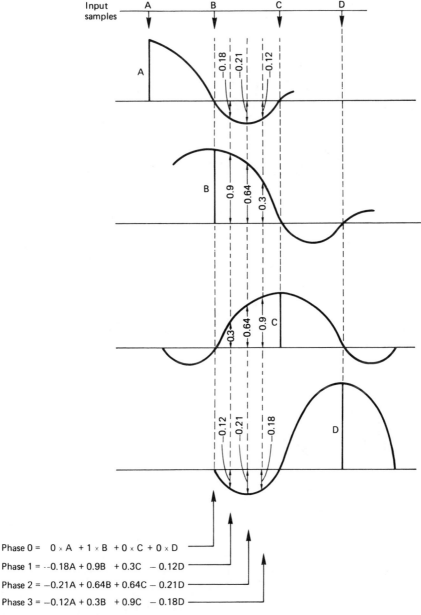

Phase 0 =   0 × A + 1 × B + 0 × C + 0 × D

Phase 1 = --0.18A + 0.9B + 0.3C − 0.12D

Phase 2 = −0.21A + 0.64B + 0.64C − 0.21D

Phase 3 = −0.12A + 0.3B + 0.9C − 0.18D

Figure 3.46 In 4× oversampling, for each set of input samples, four phases of coefficients are necessary, each of which produces one of the oversampled values.

will be replaced by a value for the nearest available filter phase. The number of phases in the filter therefore determines the accuracy of the interpolation. The effects of calculating a value for the wrong point are identical to those of sampling with clock jitter, in that an error occurs proportional to the slope of the

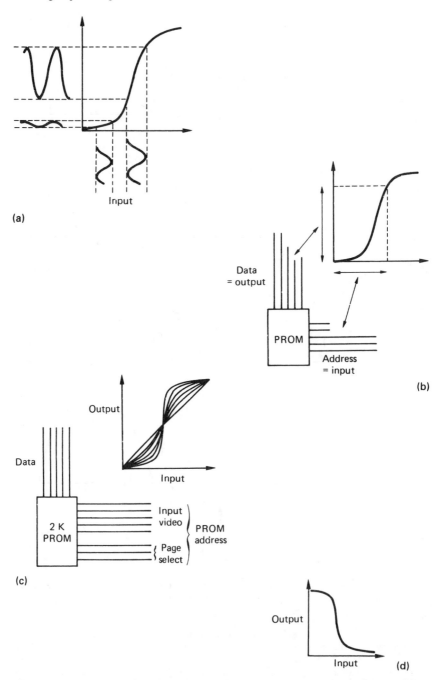

Figure 3.47 Solarization. (a) The non-linear transfer function emphasizes contrast in part of the range but reduces it at other parts. (b) The desired transfer function is implemented in a PROM. Each input sample value is used as the address to select a corresponding output value stored in the PROM. (c) A family of transfer functions can be accommodated in a larger PROM. Page select affects the high-order address bits. (d) Transfer function for luminance reversal.

signal. The result is program-modulated noise. The higher the noise specification, the greater the desired time accuracy and the greater the number of phases required. The number of phases is equal to the number of sets of coefficients available, and should not be confused with the number of points in the filter, which is equal to the number of coefficients in a set (and the number of multiplications needed to calculate one output value).

The sampling jitter accuracy necessary for 8 bit working is measured in picoseconds. This implies that something like 32 filter phases will be required for adequate performance in an 8 bit sampling-rate converter.

3.9 Digital video effects

If a RAM of the type shown in Figure 3.24 is inserted in a digital luminance path, the result will be *solarizing* which is a form of contrast enhancement. Figure 3.47 shows that a family of transfer functions can be implemented which control the degree of contrast enhancement. When the transfer function becomes so distorted that the slope reverses, the result is *luminance reversal*, where black and white are effectively interchanged. Solarizing can also be implemented in colour difference channels to obtain *chroma solarizing*. In effects machines, the degree of solarizing may need to change smoothly so that the effect can be gradually introduced. In this case the various transfer functions will be kept in different pages of a PROM, so that the degree of solarization can be selected immediately by changing the page address of the PROM. One page will have a straight transfer function, so the effect can be turned off by selecting that page.

In the digital domain it is easy to introduce various forms of quantizing distortion to obtain special effects. Figure 3.48 shows that 8 bit luminance allows 256 different brightnesses, which to the naked eye appears to be a continuous range. If some of the low-order bits of the samples are disabled, then a smaller number of brightness values describes the range from black to white. For example, if 6 bits are disabled, only 2 bits remain, and so only four possible brightness levels can be output. This gives an effect known as *contouring* since the visual effect somewhat resembles a relief map.

When the same process is performed with colour difference signals, the result is to limit the number of possible colours in the picture, which gives an effect known as *posterizing*, since the picture appears to have been coloured by paint from pots. Solarizing, contouring and posterizing cannot be performed in the composite digital domain, owing to the presence of the subcarrier in the sample values.

Figure 3.49 shows a latch in the luminance data which is being clocked at the sampling rate. It is transparent to the signal, but if the clock to the latch is divided down by some factor n, the result will be that the same sample value will be held on the output for n clock periods, giving the video waveform a staircase characteristic. This is the horizontal component of the effect known as *mosaicing*. The vertical component is obtained by feeding the output of the latch into a line memory, which stores one horizontally mosaiced line and then repeats that line m times. As n and m can be independently controlled, the mosaic tiles can be made to be of any size, and rectangular or square at will. Clearly the mosaic circuitry must be implemented simultaneously in luminance and colour difference signal paths. It is not possible to perform mosaicing on a composite digital signal, since it will destroy the subcarrier. It is common to provide a

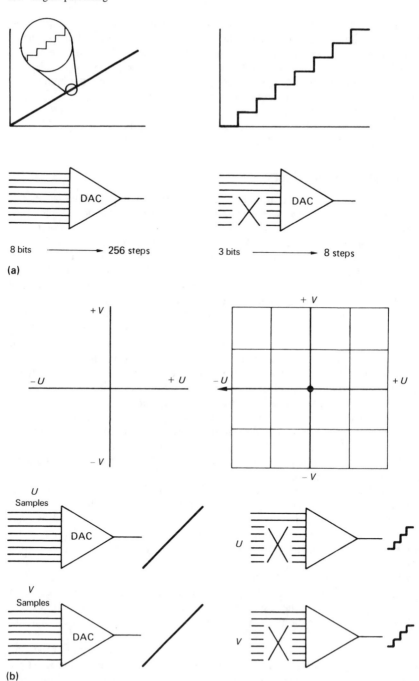

Figure 3.48 (a) In contouring, the least significant bits of the luminance samples are discarded, which reduces the number of possible output levels. (b) At left, the 8 bit colour difference signals allow 2^{16} different colours. At right, eliminating all but 2 bits of each colour difference signal allows only 2^4 different colours.

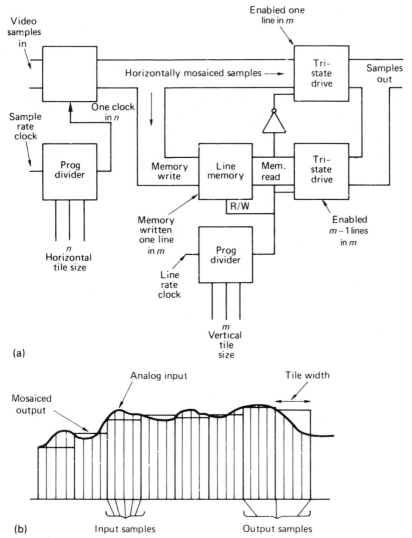

Figure 3.49 (a) Simplified diagram of mosaicing system. At left-hand side, horizontal mosaicing is done by intercepting sample clocks. On one line in m, the horizontally mosaiced line becomes the output, and is simultaneously written into a one-line memory. On the remaining $(m-1)$ lines the memory is read to produce several identical successive lines to give the vertical dimensions of the tile. (b) In mosaicing, input samples are neglected, and the output is held constant by failing to clock a latch in the data stream for several sample periods. Heavy vertical lines here correspond to the clock signal occurring. Heavy horizontal line is resultant waveform.

bypass route which allows mosaiced and unmosaiced video to be simultaneously available. Dynamic switching between the two sources controlled by a separate key signal then allows mosaicing to be restricted to certain parts of the picture.

The above simple effects did not change the shape or size of the picture, whereas the following class of manipulations will. Effects machines which manipulate video pictures are close relatives of the machines which produce computer-generated images. Computer-generated images require enormous processing power, and even with state-of-the-art CPUs the time needed to compute a single frame is too long to work with real-time video. Video effects machines, then, represent a significant technical achievement, because they have only a field period to complete the processing before another field comes along. General purpose processors are usually too slow for effects work, and most units incorporate dedicated hardware to obtain sufficient throughput.

The principle of all video manipulators is the same as the technique used by cartographers for centuries. Cartographers are faced with a continual problem in that the earth is round, and paper is flat. In order to produce flat maps, it is necessary to project the features of the round original on to a flat surface. Figure 3.50 shows an example of this. There are a number of different ways of projecting maps, and all of them must by definition produce distortion. The effect of this distortion is that distances measured near the extremities of the map appear further than they actually are. Another effect is that *great circle routes* (the shortest or longest path between two places on a planet) appear curved on a projected map. The type of projection used is usually printed somewhere on the map, a very common system being that due to Mercator. Clearly the process of mapping involves some three-dimensional geometry in order to simulate the paths of light rays from the map so that they appear to have come from the curved surface. Video effects machines work in exactly the same way.

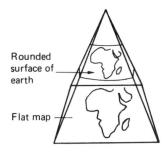

Rounded
surface of
earth

Flat map

Figure 3.50 Map projection is a close relative of video effects units which manipulate the shape of pictures.

The distortion of maps means that things are not where they seem. In timesharing computers, every user appears to have their own identical address space in which their program resides, despite the fact that many different programs are simultaneously in the memory. In order to resolve this contradiction, memory management units are constructed which add a constant value to the address which the user thinks they have (the *virtual address*) in order to produce the *physical address*. As long as the unit gives each user a different constant, they can all program in the same virtual address space without one corrupting another's programs. Because the program is no longer where it seems to be, the term mapping was introduced. The address space of a computer is one dimensional, but a video frame expressed as rows and columns of pixels can be considered to have a two-dimensional address as in

Figure 3.51. Video manipulators work by mapping the pixel addresses in two dimensions.

All manipulators must begin with an array of pixels, in which the columns must be vertical. This can only be easily obtained if the sampling rate of the incoming video is a multiple of line rate, as is done in CCIR-601. Composite video cannot be used directly, because the phase of the subcarrier would become meaningless after manipulation. If a composite video input is used, it must be decoded to baseband luminance and colour difference, so that in actuality there are three superimposed arrays of samples to be processed, one luminance and two colour difference. Most modern DVEs use CCIR-601 sampling, so digital signals conforming to this standard could be used directly from, for example, a DVTR or a hard-disk recorder.

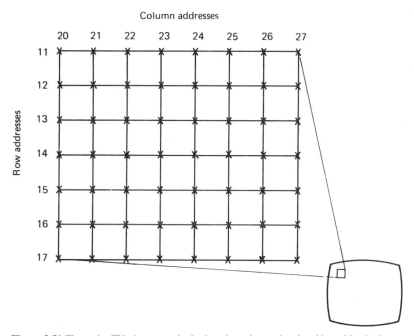

Figure 3.51 The entire TV picture can be broken down into uniquely addressable pixels.

There are two main types of effects machines, those which are field based and those which are frame based. The former are more cost conscious, the latter are more quality conscious.

Figure 3.52(a) shows an example of a field-based manipulator. In this example, the size of the picture is to be doubled. The information from lines on a given input field will appear on every other output line, and the lines in between will have to be produced by interpolation. In Figure 3.52(b) a frame-based machine is performing the same operation. Here, the lines necessary for the magnified output field come from the other input field. Clearly there is an improvement in vertical resolution to be gained by using frame-based machines, but it will be seen that there is a considerable increase in complexity.

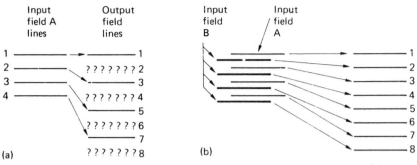

Figure 3.52 (a) In a field-based machine, interpolation becomes necessary when size of the picture is increased. (b) In a frame-based machine, the presence of the second input field allows greater vertical resolution.

A frame-based machine must produce a de-interlaced frame from which every output field can choose pixels. Producing frames at field rate sounds rather like getting something for nothing, but this is not exactly what happens. In practice, at a given field time, the lines of that field will be available, and between them will be placed the lines from the previous (most recent) field. Figure 3.53 shows that at each new field, a new pair of fields will be combined to make a frame. Unfortunately it is not possible to use the information from two different fields directly as this results in double images on moving objects. Frame-based machines must use motion sensing so that de-interlacing can be disabled when movement occurs, and interpolation used instead. More recently frame de-interlacing can be motion compensated so that full resolution remains available in moving areas. Motion compensation is considered later in this chapter. Clearly the additional cost and complexity of a frame-based processor is not trivial, but this is justified for post production work, where the utmost quality is demanded. Where an effects unit is used to increase the range of wipe patterns in an on-air video switcher, a field-based system may be acceptable because a wipe is by definition transient in nature, and there is little opportunity to study the image during the wipe.

The effect of de-interlacing is to produce an array of pixels which are the input data for the processor. In 13.5 MHz systems, the array will be about 720 pixels across, and 600 down for 50 Hz systems (500 down for 60 Hz systems). Most machines are set to strip out VITC or teletext data from the input, so that pixels representing a genuine picture appear surrounded by blanking level.

Every pixel in the frame array has an address. The address is two-dimensional, because in order to specify one pixel uniquely, the column address and the row address must be supplied. It is possible to transform a picture by simultaneously

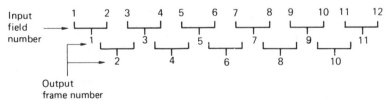

Figure 3.53 Frames are produced at field rate by combining the latest pair of fields.

addressing in rows and columns, but this is complicated, and very difficult to do in real time. It was discovered some time ago in connection with computer graphics that the two-dimensional problem can be converted with care into two one dimensional problems.[8] Essentially if a horizontal transform affecting whole rows of pixels independently of other rows is performed on the array, followed by or preceded by a vertical transform which affects entire columns independently of other columns, the effect will be the same as if a two-dimensional transform had been performed. This is the principle of separability. From an academic standpoint, it does not matter which transform is performed first. In a world which is wedded to the horizontally scanned television set, there are practical matters to consider. In order to convert a horizontal raster input into a vertical column format, a memory is needed where the input signal is written in rows but the output is read as columns of pixels.

The process of writing rows and reading columns in a memory is called transposition. Clearly two stages of transposition are necessary to return to a horizontal raster output, as shown in Figure 3.54. The vertical transform must take place between the two transposes, but the horizontal transform could take place before the first transpose or after the second one. In practice the horizontal transform cannot be placed before the first transpose, because it would interfere with the de-interlace and motion sensing process. The horizontal transform is placed after the second transpose and reads rows from the second transpose memory. As the output of the machine must be a horizontal raster, the horizontal transform can be made to work in synchronism with reference H-sync, so that the digital output samples from the H-transform can be taken direct to a DAC to become analog video with standard structure again. A further advantage is that the real-time horizontal output is one field at a time. The preceding vertical transform need only compute array values which lie in the next field to be needed at the output.

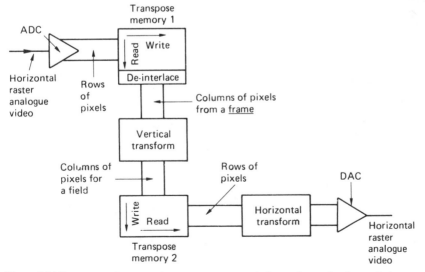

Figure 3.54 Two transposing memories are necessary, one before and one after the vertical transform.

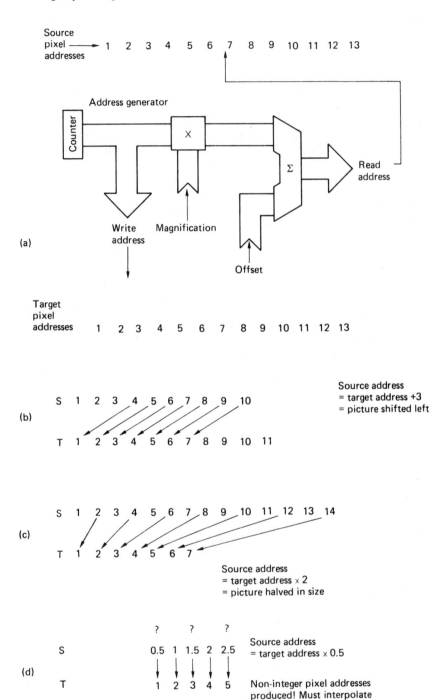

Figure 3.55 Address generation is the fundamental process behind transforms.

It is not possible to take a complete column from a memory until all of the rows have entered. The presence of the two transposes in a DVE results in an unavoidable delay of one frame in the output image. Some DVEs can manage greater delay. The effect on lip-sync may have to be considered. It is not advisable to cut from the input of a DVE to the output, since a frame will be lost, and on cutting back a frame will be repeated.

Address mapping is used to perform transforms. Now that rows and columns are processed individually, the mapping process becomes much easier to understand. Figure 3.55 shows a single row of pixels which are held in a buffer where each can be addressed individually and transferred to another. If a constant is added to the read address, the selected pixel will be to the right of the place where it will be put. This has the effect of moving the picture to the left. If the buffer represented a column of pixels, the picture would be moved vertically. As these two transforms can be controlled independently, the picture could be moved diagonally.

If the read address is multiplied by a constant, say 2, the effect is to bring samples from the input closer together on the output, so that the picture size is reduced. Again independent control of the horizontal and vertical transforms is possible, so that the aspect ratio of the picture can be modified. This is very useful for telecine work when CinemaScope films are to be broadcast. Clearly the secret of these manipulations is in the constants fed to the address generators. The added constant represents displacement, and the multiplied constant represents magnification. A multiplier constant of less than one will result in the picture getting larger. Figure 3.55 also shows, however, that there is a problem. If a constant of 0.5 is used, to make the picture twice as big, half of the addresses generated are not integers. A memory does not understand an address of two and a half! If an arbitrary magnification is used, nearly all of the addresses generated are non-integer. A similar problem crops up if a constant of less than one is added to the address in an attempt to move the picture less than the pixel spacing. The solution to the problem is interpolation. Because the input image is spatially sampled, those samples contain enough information to represent the brightness and colour all over the screen. When the address generator comes up with an address of 2.5, it actually means that what is wanted is the value of the signal interpolated half-way between pixel two and pixel three. The output of the address generator will thus be split into two parts. The integer part will become the memory address, and the fractional part is the phase of the necessary interpolation. In order to interpolate pixel values a digital filter is necessary.

Figure 3.56 shows that the input and output of an effects machine must be at standard sampling rates to allow digital interchange with other equipment. When the size of a picture is changed, this causes the pixels in the picture to fail to register with output pixel spacing. The problem is exactly the same as sampling-rate conversion, which produces a differently spaced set of samples which still represent the original waveform. One pixel value actually represents the peak brightness of a two-dimensional intensity function, which is the effect of the modulation transfer function of the system on an infinitely small point. As each dimension can be treated separately, the equivalent in one axis is that the pixel value represents the peak value of an infinitely short impulse which has been low-pass filtered to the system bandwidth. The waveform is that of a $\sin x/x$ curve, which has value everywhere except at the centre of other pixels. In order to compute an interpolated value, it is necessary to add together the contribution

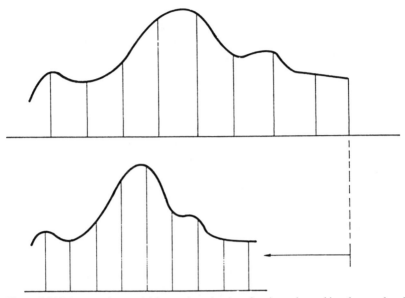

Figure 3.56 It is easy, almost trivial, to reduce the size of a picture by pushing the samples closer together, but this is not often of use, because it changes the sampling rate in proportion to the compression. Where a standard sampling-rate output is needed, interpolation must be used.

from all relevant samples, at the point of interest. Each contribution can be obtained by looking up the value of a unity $\sin x/x$ curve at the distance from the input pixel to the output pixel to obtain a coefficient, and multiplying the input pixel value by that coefficient. The process of taking several pixel values, multiplying each by a different coefficient and summing the products can be performed by the FIR (finite impulse response) configuration described earlier. The impulse response of the filter necessary depends on the magnification. Where the picture is being enlarged, the impulse response can be the same as at normal size, but as the size is reduced, the impulse response has to become broader (corresponding to a reduced spatial frequency response) so that more input samples are averaged together to prevent aliasing. The coefficient store will need a two-dimensional structure, such that the magnification and the interpolation phase must both be supplied to obtain a set of coefficients. The magnification can easily be obtained by comparing successive outputs from the address generator.

The number of points in the filter is a compromise between cost and performance, eight being a typical number for high quality. As there are two transform processes in series, every output pixel will be the result of 16 multiplications, so there will be 216 million multiplications per second taking place in the luminance channel alone for a 13.5 MHz sampling rate unit. The quality of the output video also depends on the number of different interpolation phases available between pixels. The address generator may compute fractional addresses to any accuracy, but these will be rounded off to the nearest available phase in the digital filter. The effect is that the output pixel value provided is actually the value a tiny distance away, and has the same result as sampling-clock jitter, which is to produce program-modulated noise. The greater the number of phases provided, the larger will be the size of the coefficient store needed. As the coefficient store is two-dimensional, an increase in the number of filter points and phases causes an exponential growth in size and cost. The filter itself can be implemented readily with fast multiplier chips, but one problem is accessing the memory to provide input samples. What the memory must do is to take the integer part of the address generator output and provide simultaneously as many adjacent pixels as there are points in the filter. This problem is usually solved by making the memory from several smaller memories with an interleaved address structure, so that several pixel values can be provided from one address.

In order to follow the operation of a true perspective machine, some knowledge of perspective is necessary. Stated briefly, the phenomenon of perspective is due to the angle subtended at the eye by objects being a function not only of their size but also of their distance. Figure 3.57 shows that the size of an image on the rear wall of a pinhole camera can be increased either by making the object larger or bringing it closer. In the absence of stereoscopic vision, it is not possible to tell which has happened. The pinhole camera is very useful for study of perspective, and has indeed been used by artists for that purpose. The clinically precise perspective of Canaletto paintings was achieved through the use of the camera obscura (darkened room in Latin).[9]

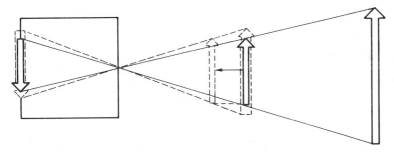

Figure 3.57 The image on the rear of the pinhole camera is identical for the two solid objects shown because the size of the object is proportional to distance, and the subtended angle remains the same. The image can be made larger (dotted) by making the object larger or moving it closer.

It is sometimes claimed that the focal length of the lens used on a camera changes the perspective of a picture. This is not true: perspective is only a function of the relative positions of the camera and the subject. Fitting a wide-angle lens simply allows the camera to come near enough to keep dramatic perspective within the frame, whereas fitting a long-focus lens allows the camera to be far enough away to display a reasonable-sized image with flat perspective.[10]

Since a single eye cannot tell distance unaided, all current effects machines work by simply producing the correct subtended angles which the brain perceives as a three-dimensional effect. Figure 3.58 shows that to a single eye, there is no difference between a three-dimensional scene and a two-dimensional image formed where rays traced from features to the eye intersect an imaginary plane. This is exactly the reverse of the map projection shown in Figure 3.50, and is the principle of all perspective manipulators.

The case of perspective rotation of a plane source will be discussed first. Figure 3.58 shows that when a plane input frame is rotated about a horizontal axis, the distance from the top of the picture to the eye is no longer the same as the distance from the bottom of the picture to the eye. The result is that the top and bottom edges of the picture subtend different angles to the eye, and where the rays cross the target plane, the image has become trapezoidal. There is now no such thing as the magnification of the picture. The magnification changes continuously from top to bottom of the picture, and if a uniform grid is input, after a perspective rotation it will appear non-linear as the diagram shows.

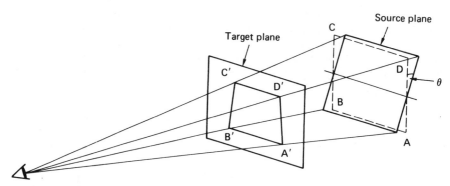

Figure 3.58 In a planar rotation effect the source plane ABCD is the rectangular input picture. If it is rotated through the angle θ, rays tracing to a single eye at left will produce a trapezoidal image A'B'C'D' on the target. Magnification will now vary with position on the picture.

The basic mechanism of the transform process has been described, but this is only half of the story, because these transforms have to be controlled. There is a lot of complex geometrical calculation necessary to perform even the simplest effect, and the operator cannot be expected to calculate directly the parameters required for the transforms. All effects machines require a computer of some kind, with which the operator communicates using keyboard entry or joystick/trackball movements at high level. These high-level commands will specify such things as the position of the axis of rotation of the picture relative to the viewer, the position of the axis of rotation relative to the source picture, and the angle of rotation in the three axes.

An essential feature of this kind of effects machine is fluid movement of the source picture as the effect proceeds. If the source picture is to be made to move smoothly, then clearly the transform parameters will be different in each field. The operator cannot be expected to input the source position for every field, because this would be an enormous task. Additionally, storing the effect would require a lot of space. The solution is for the operator to specify the picture position at strategic points during the effect, and then digital filters are used to compute the intermediate positions so that every field will have different parameters.

The specified positions are referred to as knots, nodes or keyframes, the first being the computer graphics term. The operator is free to enter knots anywhere in the effect, and so they will not necessarily be evenly spaced in time, i.e. there may well be different numbers of fields between each knot. In this environment it is not possible to use conventional FIR-type digital filtering, because a fixed-impulse response is inappropriate for irregularly spaced samples.

Interpolation of various orders is used ranging from zero-order hold for special jerky effects through linear interpolation to cubic interpolation for very smooth motion. The algorithms used to perform the interpolation are known as splines, a term which has come down from shipbuilding via computer graphics.[11] When a ship is designed, the draughtsman produces hull cross-sections at intervals along the keel, whereas the shipyard needs to recreate a continuous structure. The solution was a lead-filled bar, known as a spline, which could be formed to join up each cross-section in a smooth curve, and then used as a template to form the hull plating.

The filter which does not ring cannot be made, and so the use of spline algorithms for smooth motion sometimes results in unintentional overshoots of the picture position. This can be overcome by modifying the filtering algorithm. Spline algorithms usually look ahead beyond the next knot in order to compute the degree of curvature in the graph of the parameter against time. If a break is put in that parameter at a given knot, the spline algorithm is prevented from looking ahead, and no overshoot will occur. In practice the effect is created and run without breaks, and then breaks are added later where they are subjectively thought necessary.

It will be seen that there are several levels of control in an effects machine. At the highest level, the operator can create, store and edit knots, and specify the times which elapse between them. The next level is for the knots to be interpolated by spline algorithms to produce parameters for every field in the effect. The field frequency parameters are then used as the inputs to the geometrical computation of transform parameters which the lowest level of the machine will use as microinstructions to act upon the pixel data. Each of these layers will often have a seperate processor, not just for speed, but also to allow software to be updated at certain levels without disturbing others.

3.10 Conventional standards conversion

The interchange of video program material between line standards has been necessary because of the unfortunate number of different broadcast standards in use throughout the world. This section considers conventional standards converters which interpolate along the time axis. Motion-compensated standards conversion is treated later.

Frame- or field-based systems sample the changes of the image with respect to time. The input scene should not then change more often than one cycle per two samples, or aliasing will take place. In practice the camera has no control over the scene, and aliasing is inevitable on fast-moving objects such as helicopter rotors and stagecoach wheels, and for less obvious reasons on slow-moving fine detail. The origin of temporal frequencies due to motion is discussed in Section 3.9.

As was explained in Chapter 2, once a sampled system has aliased, the effects can never be removed, because the alias frequencies cannot be distinguished from genuine information at the same frequency. Conventional standards converters have great difficulty with aliased signals on the time axis, hence the development of motion compensation.

The bandwidth required by a television signal is obtained by multiplying the number of cycles of resolution in a line by the number of lines in a frame, and then by the frame rate. If similar vertical and horizontal resolution is assumed, bandwidth increases as the square of the resolution for a given frame rate. The adoption of interlace was claimed to allow the apparent picture rate to be higher than the frame rate. The apparent picture rate, or field rate, determines the visibility of flicker. With interlace, sending half the picture twice as often allows the bandwidth to be halved without producing flicker. Since only half of the necessary information is sent, information theory alone will tell us not to expect great things. Interlace is a form of temporal oversampling at the expense of spatial resolution. The presence of interlace makes standards conversion doubly difficult.

Figure 3.59 shows that a television signal in a certain line standard represents a series of two-dimensional images. A different line standard could have a different number of lines and a different number of frames per second. For colour broadcasts, the modulation method used to carry the colour information could also be different. The task of the standards converter is to transfer the moving image from one signal structure to the other as faithfully as possible.

Conversion from one colour modulation system to another is relatively easy. Another happy chance is that the line periods of 525 and 625 line systems are so similar that they can be assumed identical. Conversion between different numbers of lines per frame in progressive scan is relatively straightforward, but interlace makes this and conversion between different frame rates very difficult.

The handling of colour information in standards conversion is generally to decode the composite input to form separate luminance and colour difference signal channels. These are then passed to three similar baseband standards converters, although the performance of the colour difference channels would not need to be so high. Following the conversion, the output signals would then be re-encoded into the appropriate composite form if required. This approach is necessary because the transforms needed for spatial and temporal interpolation of the image would destroy the phase and thus the meaning of any subcarrier. The decoding can be considered separately from the conversion process. As there are many different types of colour modulation methods, a multistandard converter will necessarily be complex. The colour decoding and encoding techniques may be analog or digital, and the digital techniques have already been discussed in this chapter. As the composite decoding takes place prior to the standards conversion proper, it is relatively straightforward to insert a digital signal according to CCIR-601 which bypasses the composite decoder and the analog-to-digital converters. It is also feasible to incorporate the composite digital to 4:2:2 decoder to allow the input section to be entirely digital. The digital output of the standards conversion process could then be output according to CCIR-601, but in the alternative line standard. Clearly in this case the internal sampling rate of the machine will need to be 13.5 MHz.

The line period of 625/50 is almost identical to that of 525/60, and the consequence of assuming them to be exactly the same is a minute change of

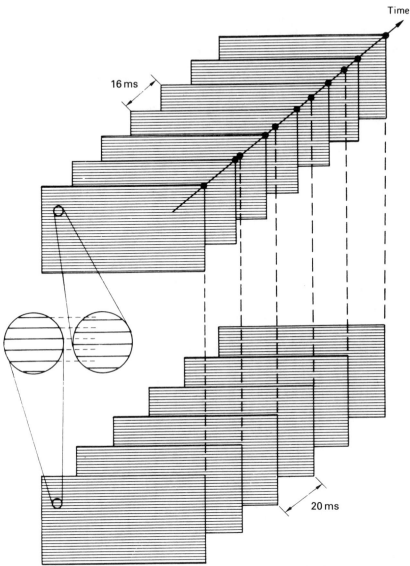

Figure 3.59 The basic problem of standards conversion. The line spacing is easy to accommodate. The field spacing is not.

horizontal magnification which is often neglected. The number of lines in a field must be changed, and the number of fields per second must be changed. In the presence of interlace, these processes must be simultaneous, but this explanation will begin with a one-dimensional approach.

In order to change the number of lines in a field, interpolation is necessary in a vertical direction. The difference in line spacing between PAL and NTSC is such that there are 21 different relationships between the lines. It is not possible to throw away lines or repeat lines, as this results in steps on diagonal edges, as shown in Figure 3.60. The figure also shows that columns of pixels are fed to a digital filter which has a low-pass response cutting off at the vertical spatial frequency of half the line spatial frequency. The necessary phase-linear characteristic is supplied by using the FIR (finite-impulse response) structure which has been described. Because there are many different line relationships, a different set of coefficients corresponding to a different offset of the impulse is necessary for each phase of interpolation.

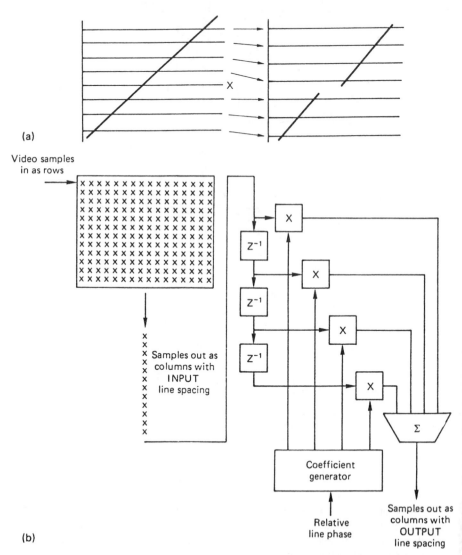

Figure 3.60 (a) Omitting or repeating lines in standards conversion causes discontinuities on diagonal detail. (b) The conversion of line spacing is by vertical FIR filtering on columns of pixels.

In a standards converter, the magnification is fixed by the relationship between the number of active lines in the two standards. Production of the best vertical resolution requires information from both fields of the input frames. Figure 3.61 shows that it is possible to supply alternate points in the FIR filter from columns of pixels in alternate fields. In this case the vertical frequency of the input is doubled, and the impulse response of the filter is sharpened in order to obtain maximum vertical resolution. Unfortunately the presence of movement in the images means that if successive fields are superimposed a double image results. It is necessary to use motion detection by comparing successive frames as has been described. When there is motion in the area of the picture being processed,

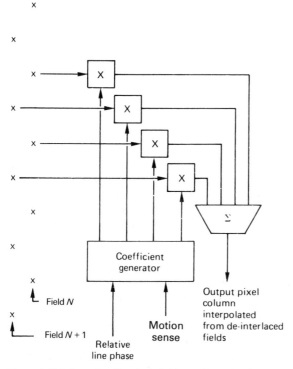

Figure 3.61 Information from both fields can be used to improve vertical resolution when there is little motion. When motion occurs, the filter impulse response must broaden, as samples from one of the fields will be neglected, and vertical resolution falls.

the pixels from the field farthest away in time are neglected by setting the coefficients fed to every other filter point to zero. The impulse response of the filter must also be broadened to account for the lack of resolution on the input. The coefficient store will thus have a two-dimensional structure, where one dimension is the repeating vertical phase relationship of the lines and the other is the amount of motion detected.[12]

As has been noted, the conversion of field rate is difficult. Figure 3.62 shows that it is not possible to omit or repeat frames because this results in

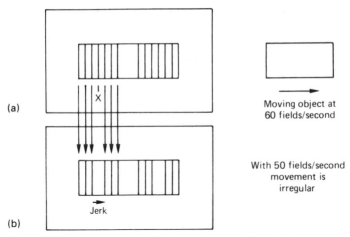

(a)

Moving object at
60 fields/second

With 50 fields/second
movement is
irregular

(b)

Figure 3.62 In (a) a rectangular object is moving smoothly to the right, and seven superimposed fields are shown in the frame in a 60 Hz system. If an attempt is made to convert to 50 Hz by omitting fields, as in (b), there will be jerky movement where the image moves a greater distance due to the omission. This is subjectively highly disturbing.

jerky movement. It is again necessary to interpolate, this time temporally. In temporal interpolation, pixels from the same physical position in several frames form the input to an FIR filter which computes what the pixel value in between frames would have been. The temporal frequency response of the filter must be one-half of the frame rate of the slower standard. Ideally the filter should have a frequency response which is flat, and then falls steeply at the cut-off frequency. In FIR filters this ideal can only be realized with a large number of filter points, and it must be appreciated that for every point in the filter, there must be a field of video stored in the machine, and an additional fast multiplier. Even the use of four fields, which is increasingly common, is less than perfect, because an FIR filter with such a small number of points must compromise between the conflicting requirements of stopband rejection and rate of roll-off. As the filter is designed to suppress the unwanted field rate in the information spectrum, stopband rejection is important. The symptom of insufficient stopband rejection would be beats between the two field rates modulating the brightness. Since this is highly objectionable, the rate of roll-off is generally sacrificed instead, which means that the temporal frequency response of the converter is impaired.

Figure 3.63 shows that the input to a two-dimensional filter can be considered to be pixels in columns whose height represents distance up the tube face, spaced apart by the period between fields. Owing to interlace, the pixels in odd fields have a vertical displacement relative to the pixels in even fields. Assuming a typical value of four filter points in each dimension, four pixels from four fields are combined to produce one output pixel by providing each input pixel with a coefficient by which it is multiplied before adding the products. The output pixel will have a temporal and spatial phase which determines what the coefficients must be.

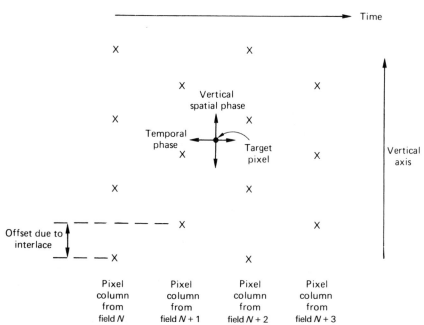

Figure 3.63(a) In a two-dimensional digital filter, columns from several fields form the input, and the target pixel phase vertically and temporally controls the coefficients. The structure of the filter is exactly that of an FIR filter, only the sources of information cause the two-dimensional effect.

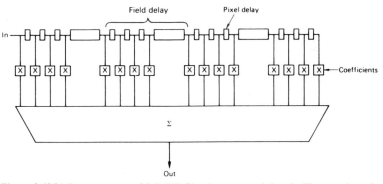

Figure 3.63(b) Rearrangement of 2-D FIR filter into temporal domain illustrates how the impulse response spreads over many fields because of the low temporal frequency response of television systems. Note that input is pixel columns.

Figure 3.64 shows that if the temporal impulse response is considered first, an interpolated pixel value is obtained by computing the time at which the pixel is needed in relation to the time of available fields. In 60 Hz to 50 Hz conversion, input spacing will be 16.7 milliseconds and output spacing will be 20 milliseconds. The time determines the filter phase, which has the effect of shifting the impulse response of the filter through time to allow a greater contribution from near fields and a smaller contribution from far fields.

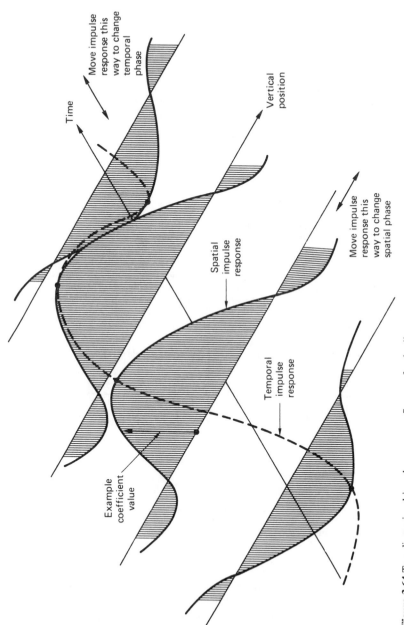

Figure 3.64 Two-dimensional impulse response. See text for details.

The vertical phase of the desired pixel in relation to the available pixels will then determine the vertical phase of the filter. The sum of the four coefficients used in a given field column will equal the (imaginary) coefficient of a one-dimensional temporal filter at that field.

Where the difference between successive frames is very small, this can be assumed to be due to random noise, and broadening the impulse response has the effect of reducing the noise.

It is not practicable to optimize the action of such a device for all material, because good performance on noisy inputs may result in unnatural removal of rain and reflections on water in normal video inputs. It is usual to provide an operator control so that the motion sensing can be made more or less active according to the subject matter.

Where the difference between pixels in successive frames is large, this must be due to motion, and it will be necessary to modify the two-dimensional impulse response to prevent multiple images. Essentially what happens is that the coefficients for fields which contain movement are reduced, so their contribution to the output pixel becomes smaller, and the coefficients fed to remaining fields increase to compensate.

3.11 Motion and resolution

The human eye resembles a CCD camera in that the retina is covered with a large number of discrete sensors which are used to build up an image. The spacing between the sensors has a similar effect on the resolution of the eye to the number of pixels in a CCD chip. However, the eye acts to a degree as if it were AC coupled so that its response to low spatial frequencies (the average brightness of a scene) falls.

The eye also has a temporal response taking the form of a lag known as persistence of vision. The effect of the lag is that resolution is lost in areas where the image is moving rapidly with respect to the retina, a phenomenon known as motion blur.

Thus a fixed eye has poor resolution of moving objects but we are not generally aware of this because the eye can move to follow objects of interest. Figure 3.65 shows the difference this makes. In (a) a detailed object moves past a fixed eye. It does not have to move very fast before the temporal frequency at a fixed point on the retina rises beyond the temporal response of the eye and there is motion blur. In (b) the eye is following the moving object and as a result the temporal frequency at a fixed point on the retina is zero; the full resolution is then available because the image is stationary with respect to the eye. In real life we can see moving objects in some detail unless they move faster than the eye can follow.

Similar processes occur when watching television. Television viewing is affected both by the motion of the original scene with respect to the camera and by the motion of the eye with respect to the display. The situation is further complicated by the fact that television pictures are not continuous, but are sampled at the field rate. When there is relative movement between camera and scene, detailed areas develop high temporal frequencies, just as was shown in Figure 3.65 for the eye. This is because relative motion results in a given point on the camera sensor effectively scanning across the scene. The temporal frequencies generated are beyond the limit set by sampling theory, and aliasing should take place.

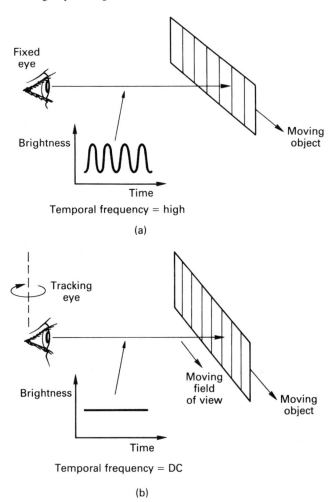

Figure 3.65 In (a) a detailed object moves past a fixed eye, causing temporal frequencies beyond the response of the eye. This is the cause of motion blur. In (b) the eye tracks the motion and the temporal frequency becomes zero. Motion blur cannot then occur.

However, when the resultant pictures are viewed by a human eye, this aliasing is not perceived because, once more, the eye attempts to follow the motion of the scene. Figure 3.66 shows what happens when the eye follows correctly. The original scene and the retina are now stationary with respect to one another, but the camera sensor and display are both moving through the field of view. As a result the temporal frequency at the eye due to the object being followed is brought to zero and no aliasing is perceived by the viewer due to the field-rate sampling.

The viewer's impression is not quite the same as if the scene were being viewed through a piece of moving glass, because of the movement of the image relative to the camera sensor and the display. Here, temporal frequencies do exist,

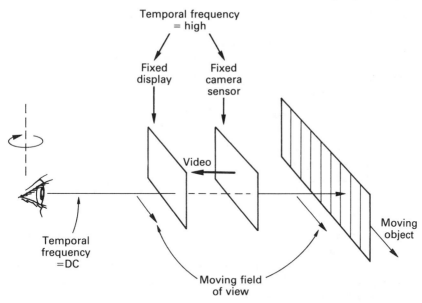

Figure 3.66 An object moves past a camera, and is tracked on a monitor by the eye. The high temporal frequencies cause aliasing in the TV signal, but these are not perceived by the tracking eye as this reduces the temporal frequency to zero. Compare with Figure 3.65.

and the temporal aperture effect (lag) of both will reduce perceived resolution. This is the reason that shutters are sometimes fitted to CCD cameras used at sporting events. The mechanically rotating shutter allows light onto the CCD sensor for only part of the field period thereby reducing the temporal aperture. The result is obvious from conventional photography in which one naturally uses a short exposure for moving subjects. The shuttered CCD camera effectively has an exposure control.

On the other hand a tube camera displays considerable lag and will not perform as well under these circumstances.

3.12 Motion-compensated standards conversion

The conventional four-field converters shown in Section 3.10 interpolate through time, using pixel data from four input fields in order to compute the likely values of pixels in an intermediate output field. This large temporal aperture is required to eliminate beat frequencies between the two field rates, but it does not handle motion transparently.

Figure 3.67(a) shows that if an object is moving, it will be in a different place in successive fields. Interpolating between several fields results in multiple images of the object. The position of the dominant image will not move smoothly, an effect which is perceived as judder. If, however, the camera is panning the moving object, it will be in much the same place in successive fields and Figure 3.67(b) shows that it will be the background which judders.

This can be overcome by using motion compensation, which can be followed by considering what happens in the spatio-temporal volume. A conventional

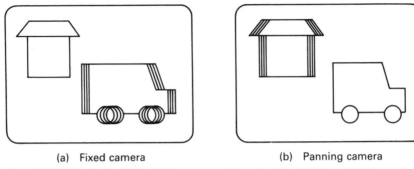

(a) Fixed camera   (b) Panning camera

Figure 3.67 (a) Conventional four-field converter with moving object suffers multiple images. (b) If the moving object is panned, the judder moves to the background.

standards converter interpolates only along the time axis, whereas a motion-compensated standards converter can swivel its interpolation axis off the time axis. Figure 3.68(a) shows the input fields in which three objects are moving in a different way. In (b) it will be seen that the interpolation axis is aligned with the trajectory of each moving object in turn.

This has a dramatic effect. Each object is no longer moving with respect to its own interpolation axis, and so on that axis it no longer generates temporal frequencies due to motion and temporal aliasing cannot occur. Interpolation along the correct axes will then result in a sequence of output fields in which motion is properly portrayed. Motion-compensated converters rely on a quite separate process which has measured the motion. Motion estimation is literally a process which analyses successive fields and determines how objects move from one to the next.

Figure 3.69 shows the sequence of events in a motion-compensated standards converter. The motion estimator measures movements between successive fields. These motions must then be attributed to objects by creating boundaries around sets of pixels having the same motion. The result of this process is a set of motion vectors, hence the term vector assignment. The motion vectors are then input to a modified four-field standards converter in order to deflect the interfield interpolation axis.

3.13 Introduction to data reduction

The data rates resulting from conventional PCM video are quite high, and although recorders have been developed which store the signals in this form, they will remain exclusively in the professional domain for the foreseeable future. For production purposes, the simplicity of PCM video is attractive because it allows manipulations with the minimum of quality loss. The signals seldom need to travel long distances and the provision of sufficient data rate is not an issue. However, the increasing range of applications of digital video are taking it out of the studio and into the consumer domain. In this case transmission over long distances is required, where the cost is generally proportional to the bit rate. There is then an obvious pressure to reduce the data rate. The running costs of a professional DVTR are reasonable in the context of the overall cost of a television production, but beyond the resources of the individual. If digital video

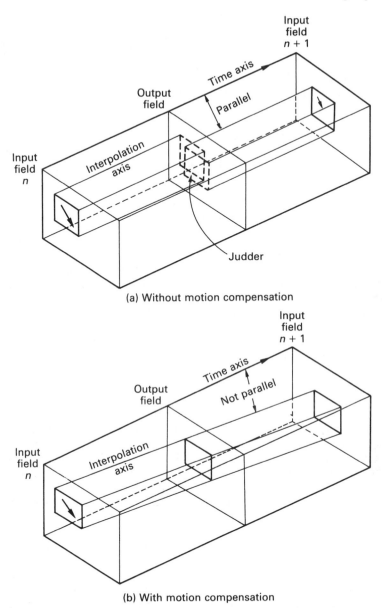

(a) Without motion compensation

(b) With motion compensation

Figure 3.68 (a) Input fields with moving object. (b) Moving the interpolation axis to make it parallel to the trajectory of the object.

recording is to be made available to the consumer, the running cost and tape consumption will need to be reduced considerably. Digital video recording in this context includes new developments such as interactive video (CD-I) which requires rapid random access to images on optical disk.

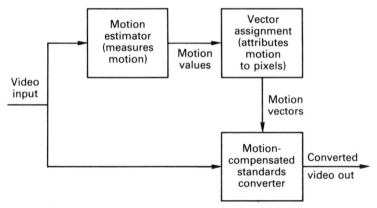

Figure 3.69 The essential stages of a motion-compensated standards converter.

As PCM high definition requires about five times the data rate of normal definition then the use of data reduction will be mandatory to deliver the signal to the home and will be considered for production recorders.

The use of data reduction also allows the life of an existing transport and tape design to be extended by raising the apparent data rate. Ampex have used data reduction in order to record 4:2:2 component data on a D-2 type transport; 4:2:2 requires approximately twice the data rate as NTSC sampled at four times subcarrier.

Data reduction is a flexible technology because the degree of coding complexity and the degree of compression used can be varied to suit the application. Video contains redundancy because typical images contain areas which are similar. The actual information in video is known as the entropy, which is the unpredictable or new part of the signal, and the remainder is redundancy, which is a part of the signal which is predictable. The sum of the two is the original data rate. The degree of compression cannot be so severe that the new data rate is less than the entropy, as information must then be lost. In theory, all of the redundancy could be removed, leaving only the entropy, but this would require a perfect algorithm which would be extremely complex. In practice the compression factor will be less than this so that some leeway is available. This allows simpler algorithms to be used and where necessary also permits multiple generations without artefacts being visible. Thus production DVTRs such as Sony's Digital Betacam and the Ampex DCT use only very mild compression of around 2:1.

For production recorders, only redundancy within the field is used, and no advantage is taken of the redundancy between fields as this would compromise editing. Clearly a consumer DVTR needs only single-generation operation and has simple editing requirements. A much greater degree of compression can then be used, which might also take advantage of redundancy between fields. The same is true for broadcasting, where bandwidth is at a premium. There is now no doubt that the future of television broadcasting (and radio for that matter) lies in data-reduced digital technology. Analog simply requires too much bandwidth.

There are thus a number of identifiable and different approaches to data reduction. For production purposes, *intrafield* data reduction is used with a mild compression factor in order to allow maximum editing freedom with negligible occurrence of artefacts. Compression algorithms intended for transmission of still images in other applications such as wirephotos can be adapted for intrafield video compression. The ISO JPEG (Joint Photographic Experts Group)[13,14] standard is such an algorithm. *Interfield* data reduction allows higher compression factors with infrequent artefacts for the delivery of post-produced material to the consumer. With even higher compression factors, leading to frequent artefacts, non-critical applications such as videophones and games are supported where the data rate has to be as low as possible. The ISO MPEG 1 (Moving Picture Experts Group)[15] standards address these applications.

3.14 DCT-based data reduction

This type of data reduction takes each individual field (or frame in progressive scan standards) and treats it in isolation from any other field or frame. The most common algorithms are based on the discrete cosine transform described in section 3.7.

Figure 3.70 shows an example of the different coefficients of a DCT for an 8×8 pixel block, and adding these together in different proportions will give any original pixel block. The top left coefficient conveys the DC component of the block. This one will be a unipolar (positive-only) value in the case of

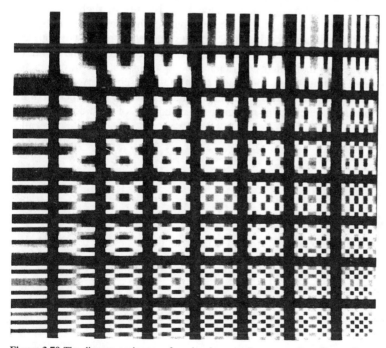

Figure 3.70 The discrete cosine transform breaks up an image area into discrete frequencies in two dimensions. The lowest frequency can be seen here at the top left corner. Horizontal frequency increases to the right and vertical frequency increases downwards.

luminance and will typically be the largest value in the block as the spectrum of typical video signals is dominated by the DC component. Moving to the right the coefficients represent increasing horizontal spatial frequencies and moving downwards the coefficients represent increasing vertical spatial frequencies. The bottom right coefficient represents the highest diagonal frequencies in the block. All of these coefficients are bipolar, where the polarity indicates whether the original spatial waveform at that frequency was inverted.

In typical pictures, the coefficients representing the higher two-dimensional spatial frequencies will be zero or of small value in large areas of typical video, due to motion blurring or simply plain undetailed areas before the camera. In general, the further from the top left corner the coefficient is, the smaller will be its magnitude on average. Coding gain (the technical term for reduction in the number of bits needed) is achieved by taking advantage of the zero and low-valued coefficients to cut down on the data necessary. Thus it is not the DCT which compresses the data, it is the subsequent processing. The DCT simply expresses the data in a form which makes the subsequent processing easier. Thus the correct terminology is to say that a compression algorithm is *DCT-based*.

Once transformed, there are various techniques which can be used to reduce the data needed to carry the coefficients. These will be based on a knowledge of the signal statistics and the human vision mechanism and will often be combined in practical systems.[16,17]

Psycho-visual knowledge will be used to process the coefficients. Omitting a coefficient means that the appropriate frequency component is missing from the reconstructed block. The difference between original and reconstructed blocks is regarded as noise added to the wanted data. The visibility of such noise is far from uniform. Figure 3.71 shows that the sensitivity of the eye to noise falls with frequency. The maximum sensitivity is at DC and as a result the top left coefficient is often treated as a special case and left unchanged. It may warrant more error protection than other coefficients.

Psycho-visual coding takes advantage of the falling sensitivity to noise by multiplying each coefficient by a different weighting constant as a function of its frequency. This has the effect of reducing the magnitude of each coefficient so

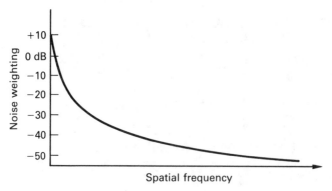

Figure 3.71 The sensitivity of the eye to noise is greatest at low frequencies and drops rapidly with increasing frequency. This can be used to mask quantizing noise caused by the compression process.

that fewer bits are needed to represent it. Another way of looking at this process is that the coefficients are individually requantized with step sizes which increase with frequency. The larger step size increases the quantizing noise at frequencies where it is not visible.

Knowledge of the signal statistics gained from extensive analysis of real material can be used to describe the probability of a given coefficient having a given value. This is the basis of entropy coding, in which coefficients are described not by fixed wordlength numbers, but by variable-length codes. The shorter codes are allocated to the most probable values and the longer codes to the least probable values. This allows a coding gain on typical signals. One of the best known variable-length codes is the Huffman code.[18]

The main difficulty with variable-length codes is separating the symbols when they are serialized. With fixed wordlength, the bit clock is merely divided by the wordlength to obtain a word clock. With variable-length coding the bit stream coding must be such that the decoder can determine the boundaries between words unaided. Figure 3.72 shows an example of Huffman coding.

Coefficient	Code	Number of zeros	Code
1	1	1	11
2	001	2	101
3	0111	3	011
4	00001	4	0101
5	01101	5	0011
6	011001	etc.	etc.
7	0000001		
Run-length code	010		

Figure 3.72 In Huffman coding the most probable coefficient values are allocated to the shortest codes. All zero coefficients are coded with run-length coding which counts the number of zeros.

When serializing a coefficient block, it is normal to scan in a sequence where the largest coefficient values are scanned first. Clearly such a scan begins in the top left corner and ends in the bottom right corner. Statistical analysis of real program material can be used to determine an optimal scan, but in many cases a regular zig-zag scan shown in Figure 3.73 will be used with slight loss of performance. The advantage of such a scan is that on typical material the scan finishes with coefficients which are zero valued. Instead of transmitting these zeros, a unique 'end of block' symbol is transmitted instead. Just before the last finite coefficient and the EOB symbol it is likely that some zero-value coefficients will be scanned. The coding enters a different mode whereby it simply transmits a unique prefix called a run-length prefix, followed by a code specifying the number of zeros which follow. This is also shown in Figure 3.72.

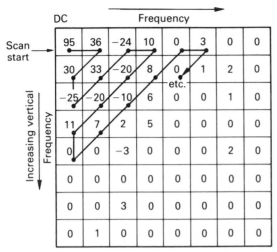

Figure 3.73 The zig-zag scan starting top left orders the coefficients in the best sequence for compression as the later ones will have smaller value.

Figure 3.74(a) shows a block diagram of a representative image data reduction unit. The input image is blocked, and the DCT stage transforms the blocks into a form in which redundancy can be identified. Psycho-visual weighting then reduces coefficient values according to the human visual process. Block scanning and variable-length/run-length coding finish the job. The receiver is shown in Figure 3.74(b). The input bit stream is deserialized into symbols, and the run-length decoder reassembles the runs of zeros. The variable-length decoder then converts back to constant wordlength coefficients. The psycho-visual weighting is reversed by a division which cancels the original multiplication. An inverse DCT then reconstructs the blocks.

3.15 Motion-compensated data reduction

Higher compression factors are easier to obtain if advantage is taken of redundancy between successive images. Only the difference between images need be sent. Clearly with a still picture, successive images will be identical and the difference will be trivial. In practice, movement reduces the similarity between successive images and the difference data increase. One way of increasing the coding gain is to use motion compensation. If the motion of an object between images is known and transmitted, the decoder can use the motion vector to shift the pixel data describing the object in the previous image to the correct position in the current image. The image difference for the object will then be smaller. The differences between images will be compressed by a DCT-based system as described above.

Figure 3.75 shows a motion-compensated system. Incoming video passes in parallel to the motion estimator and the line scan to block scan converter. The motion estimator compares the incoming frame with the previous one in the frame store in order to measure motion, and sends the motion vectors to the motion compensation unit which shifts objects in the frame store output to their

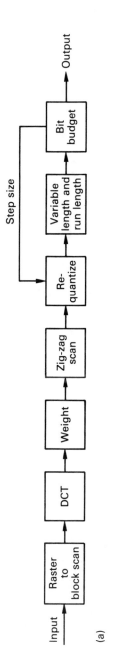

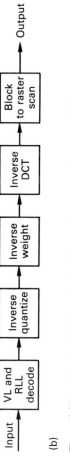

Figure 3.74 An intrafield DCT-based coder (a) and the corresponding decoder (b).

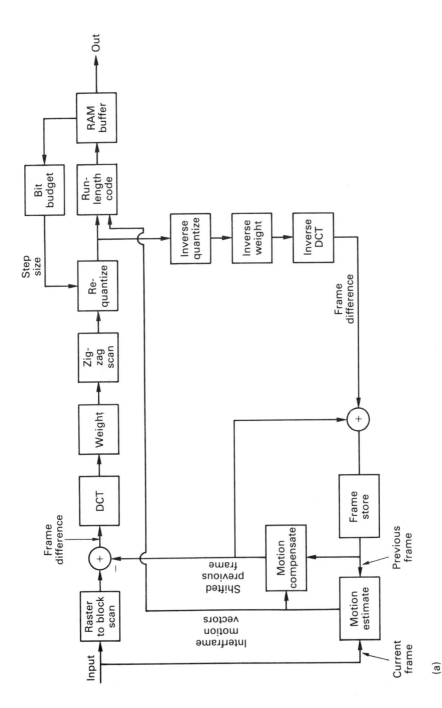

Figure 3.75 An interfield motion-compensated coder (a) and the corresponding decoder (b). See text for details.

(a)

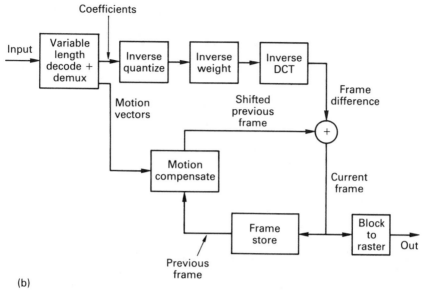

(b)

Figure 3.75 (continued) .

estimated positions in the new frame. This results in a predicted frame, which is subtracted from the input frame in order to obtain the frame difference or prediction error. The result of this process is that the temporal or interframe redundancy and the entropy have been separated. The redundancy is the predictable part of the image as anything which is predictable carries no information. The prediction error is the entropy, the part which could not be predicted. The frame difference is then processed to remove spatial redundancy. This is done with a combination of DCT, weighting and quantizing. The spatially reduced frame difference is combined with the motion vectors in order to produce the system output.

It will be seen from Figure 3.75(a) that there is also a local decoder which consists of an inverse quantizer, inverse weighting stage and an inverse DCT. Adding the locally decoded prediction error (image difference) to the predicted frame must result in the original frame (plus quantizing noise) which updates the frame store. Figure 3.75(b) shows the decoder. The frame store output is shifted by the transmitted motion vectors and the result is the same predicted frame as was produced in the encoder. The decoded frame error is added to the predicted frame and the original frame results.

References

1. RAY, S.F., *Applied photographic optics*. Oxford: Focal Press, Ch. 17 (1988)
2. KRANIAUSKAS, P., *Transforms in signals and systems*. Wokingham: Addison Wesley, Ch. 6 (1992)
3. AHMED, N., NATARAJAN, T. and RAO, K., Discrete Cosine Transform, *IEEE Trans. Computers*, **C-23**, 90–93 (1974)
4. DE WITH, P.H.N., Data compression techniques for digital video recording. Ph.D. Thesis, Technical University of Delft (1992)

5. VAN DEN ENDEN, A.W.M. and VERHOECKX, N.A.M., Digital signal processing: theoretical background. *Philips Tech. Rev.*, **42**, 110–144 (1985)

6. MCCLELLAN, J.H., PARKS, T.W. and RABINER, L.R., A computer program for designing optimum FIR linear-phase digital filters. *IEEE Trans. Audio Electroacoust.* **AU-21**, 506–526 (1973)

7. CROCHIERE, R.E. and RABINER, L.R., Interpolation and decimation of digital signals – a tutorial review. *Proc. IEEE*, **69**, 300–331 (1981)

8. NEWMAN, W.M. and SPROULL, R.F., *Principles of interactive computer graphics*. Tokyo: McGraw-Hill (1979)

9. GERNSHEIM, H., *A Concise History of Photography*. London: Thames and Hudson, pp. 9–15 (1971)

10. HEDGECOE, J., *The Photographer's Handbook*. London: Ebury Press, pp. 104–105 (1977)

11. DE BOOR, C., *A practical guide to splines*. Berlin: Springer, (1978)

12. BEANLAND, D., ISIS. *Int. Broadcast Eng.*, **20** No. 229, 40–43 (1989)

13. ISO Joint Photographic Experts Group standard JPEG-8-R8

14. WALLACE, G.K., Overview of the JPEG (ISO/CCITT) still image compression standard. ISO/JTC1/SC2/WG8 N932 (1989)

15. LE GALL, MPEG: a video compression standard for multimedia applications. *Commun. ACM*, **34**, No. 4, 46–58 (1991)

16. CLARKE, R.J. *Transform coding of images*, New York: Academic Press (1985)

17. NETRAVALI, A.N. and HASKELL, B.G., *Digital pictures – representation and compression*, New York: Plenum Press (1988)

18. HUFFMAN, D.A. A method for the construction of minimum redundancy codes. *Proc. IRE*, **40**, 1098–1101 (1952)

Chapter 4

Digital coding principles

Recording and transmission are quite different tasks, but they have a great deal in common. Digital transmission consists of converting data into a waveform suitable for the path along which it is to be sent. Digital recording is basically the process of recording a digital transmission waveform on a suitable medium. In this chapter the fundamentals of digital recording and transmission are introduced along with descriptions of the coding and error correction techniques used in practical applications.

4.1 Introduction to the channel

Data can be recorded on many different media and conveyed using many forms of transmission. The generic term for the path down which the information is sent is the *channel*. In a transmission application, the channel may be no more than a length of cable. In a recording application the channel will include the record head, the medium and the replay head. In analog systems, the characteristics of the channel affect the signal directly. It is a fundamental strength of digital video that by pulse code modulating a waveform the quality can be made independent of the channel.

In digital circuitry there is a great deal of noise immunity because the signal has only two states, which are widely separated compared with the amplitude of noise. In both digital recording and transmission this is not always the case. In magnetic recording, noise immunity is a function of track width and reduction of the working SNR of a digital track allows the same information to be carried in a smaller area of the medium, improving economy of operation. In broadcasting, the noise immunity is a function of the transmitter power and reduction of working SNR allows lower power to be used with consequent economy. These reductions also increase the random error rate, but, as was seen in Chapter 1, an error-correction system may already be necessary in a practical system and it is simply made to work harder.

In real channels, the signal may *originate* with discrete states which change at discrete times, but the channel will treat it as an analog waveform and so it will not be *received* in the same form. Various frequency-dependent loss mechanisms will reduce the amplitude of the signal. Noise will be picked up in the channel as a result of stray electric fields or magnetic induction. As a result the voltage received at the end of the channel will have an infinitely varying state along with a degree of uncertainty due to the noise. Different frequencies can propagate at different speeds in the channel; this is the phenomenon of group delay. An alternative way of considering group delay is that there will be frequency-dependent phase shifts in the signal and these will result in uncertainty in the timing of pulses.

In digital circuitry, the signals are generally accompanied by a separate clock signal which reclocks the data to remove jitter as was shown in Chapter 1. In contrast, it is generally not feasible to provide a separate clock in recording and transmission applications. In the transmission case, a separate clock line would not only raise cost, but is impractical because at high frequency it is virtually impossible to ensure that the clock cable propagates signals at the same speed as the data cable except over short distances. In the recording case, provision of a separate clock track is impractical at high density because mechanical tolerances cause phase errors between the tracks. The result is the same: timing differences between parallel channels which are known as skew.

The solution is to use a self-clocking waveform and the generation of this is a further essential function of the coding process. Clearly if data bits are simply clocked serially from a shift register in so-called direct recording or transmission this characteristic will not be obtained. If all the data bits are the same, for example all zeros, there is no clock when they are serialized.

It is not the channel which is digital; instead the term describes the way in which the received signals are *interpreted*. When the receiver makes discrete decisions from the input waveform it attempts to reject the uncertainties in voltage and time. The technique of channel coding is one where transmitted waveforms are restricted to those which still allow the receiver to make discrete decisions despite the degradations caused by the analog nature of the channel.

4.2 Types of transmission channel

Transmission can be by electrical conductors, radio or optical fibre. Although these appear to be completely different, they are in fact just different examples of electromagnetic energy travelling from one place to another. If the energy is made to vary in some way, information can be carried.

Electromagnetic energy propagates in a manner which is a function of frequency, and our partial understanding requires it to be considered as electrons, waves or photons so that we can predict its behaviour in given circumstances.

At DC and at the low frequencies used for power distribution, electromagnetic energy is called electricity and it is remarkably aimless stuff which needs to be transported completely inside conductors. It has to have a complete circuit to flow in, and the resistance to current flow is determined by the cross-sectional area of the conductor. The insulation around the conductor and the spacing between the conductors has no effect on the ability of the conductor to pass current. At DC an inductor appears to be a short circuit, and a capacitor appears to be an open circuit.

As frequency rises, resistance is exchanged for impedance. Inductors display increasing impedance with frequency, capacitors show falling impedance. Electromagnetic energy becomes increasingly desperate to leave the conductor. The first symptom is the skin effect: the current flows only in the outside layer of the conductor, effectively causing the resistance to rise.

As the energy is starting to leave the conductors, the characteristics of the space between them become important. This determines the impedance. A change of impedance causes reflections in the energy flow and some of it heads back towards the source. Constant-impedance cables with fixed conductor spacing are necessary, and these must be suitably terminated to prevent reflections. The most important characteristic of the insulation is its thickness as this determines the spacing between the conductors.

As frequency rises still further, the energy travels less in the conductors and more in the insulation between them, and their composition becomes important and they begin to be called dielectrics. A poor dielectric like PVC absorbs high-frequency energy and attenuates the signal. So-called low-loss dielectrics such as PTFE are used, and one way of achieving low loss is to incorporate as much air in the dielectric as possible by making it in the form of a foam or extruding it with voids.

This frequency-dependent behaviour is the most important factor in deciding how best to harness electromagnetic energy flow for information transmission. It is obvious that the higher the frequency, the greater the possible information rate, but in general losses increase with frequency, and flat frequency response is elusive. The best that can be managed is that over a narrow band of frequencies, the response can be made reasonably constant with the help of equalization. Unfortunately raw data when serialized have an unconstrained spectrum. Runs of identical bits can produce frequencies much lower than the bit rate would suggest. One of the essential steps in a transmission system is to modify the spectrum of the data into something more suitable.

At moderate bit rates, say a few megabits per second, and with moderate cable lengths, say a few metres, the dominant effect will be the capacitance of the cable due to the geometry of the space between the conductors and the dielectric between. The capacitance behaves under these conditions as if it were a single capacitor connected across the signal. The effect of the series source resistance and the parallel capacitance is that signal edges or transitions are turned into exponential curves as the capacitance is effectively being charged and discharged through the source impedance.

As cable length increases, the capacitance can no longer be lumped as if it were a single unit; it has to be regarded as being distributed along the cable. With rising frequency, the cable inductance also becomes significant, and it too is distributed.

The cable is now a transmission line and pulses travel down it as current loops which roll along as shown in Figure 4.1. If the pulse is positive, as it is launched along the line, it will charge the dielectric locally as in (a). As the pulse moves along, it will continue to charge the local dielectric as in (b). When the driver finishes the pulse, the trailing edge of the pulse follows the leading edge along the line. The voltage of the dielectric charged by the leading edge of the pulse is now higher than the voltage on the line, and so the dielectric discharges into the line as in (c). The current flows forward as it is in fact the same current which is flowing into the dielectric at the leading edge. There is thus a loop of current rolling down the line flowing forwards in the 'hot' wire and backwards in the return.

The constant to-ing and fro-ing of charge in the dielectric results in dielectric loss of signal energy. Dielectric loss increases with frequency and so a long transmission line acts as a filter. Thus the term 'low-loss' cable refers primarily to the kind of dielectric used.

Transmission lines which transport energy in this way have a characteristic impedance caused by the interplay of the inductance along the conductors with

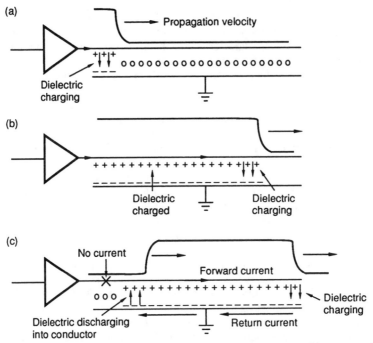

Figure 4.1 A transmission line conveys energy packets which appear with respect to the dielectric. In (a) the driver launches a pulse which charges the dielectric at the beginning of the line. As it propagates the dielectric is charged further along as in (b). When the driver ends the pulse, the charged dielectric discharges into the line. A current loop is formed where the current in the return loop flows in the opposite direction to the current in the 'hot' wire.

the parallel capacitance. One consequence of that transmission mode is that correct termination or matching is required between the line and both the driver and the receiver. When a line is correctly matched, the rolling energy rolls straight out of the line into the load and the maximum energy is available. If the impedance presented by the load is incorrect, there will be reflections from the mismatch. An open circuit will reflect all of the energy back in the same polarity as the original, whereas a short circuit will reflect all of the energy back in the opposite polarity. Thus impedances above or below the correct value will have a tendency towards reflections whose magnitude depends upon the degree of mismatch and whose polarity depends upon whether the load is too high or too low. In practice it is the need to avoid reflections which is the most important reason to terminate correctly.

A perfectly square pulse contains an indefinite series of harmonics, but the higher ones suffer progressively more loss. A square pulse at the driver becomes less and less square with distance as Figure 4.2 shows. The harmonics are progressively lost until in the extreme case all that is left is the fundamental. A transmitted square wave is received as a sine wave. Fortunately data can still be recovered from the fundamental signal component.

Once all the harmonics have been lost, further losses cause the amplitude of the fundamental to fall. The effect worsens with distance and it is necessary to ensure that data recovery is still possible from a signal of unpredictable level.

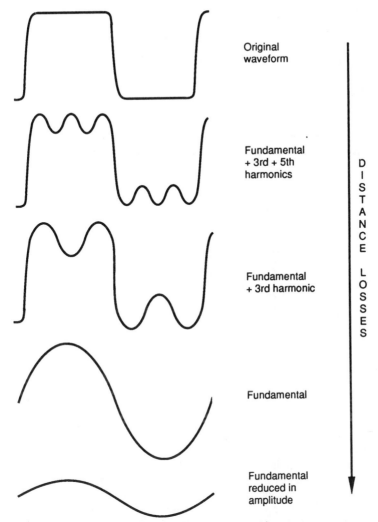

Figure 4.2 A signal may be square at the transmitter, but losses increase with frequency, and as the signal propagates, more of the harmonics are lost until only the fundamental remains. The amplitude of the fundamental then falls with further distance.

4.3 Types of recording medium

Digital media do not need to have linear transfer functions, nor do they need to be noise-free or continuous. All they need to do is to allow the player to be able to distinguish the presence or absence of replay events, such as the generation of pulses, with reasonable (rather than perfect) reliability. In a magnetic medium, the event will be a flux change from one direction of magnetization to another. In an optical medium, the event must cause the pickup to perceive a change in the intensity of the light falling on the sensor. In some disks it will be through selective absorption of light by dyes. In magneto-optical disks the recording itself is magnetic, but it is made and read using light.

4.4 Magnetic recording

Magnetic recording relies on the hysteresis of certain magnetic materials. After an applied magnetic field is removed, the material remains magnetized in the same direction. By definition the process is non-linear, and analog magnetic recorders have to use bias to linearize it. Digital recorders are not concerned with the non-linearity, and HF bias is unnecessary.

Figure 4.3 shows the construction of a typical digital record head, which is not dissimilar to an analog record head. A magnetic circuit carries a coil through which the record current passes and generates flux. A non-magnetic gap forces the flux to leave the magnetic circuit of the head and penetrate the medium. The

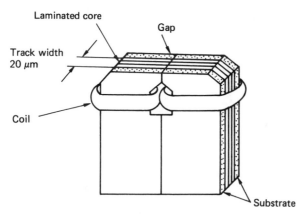

Figure 4.3 A digital record head is similar in principle to an analog head but uses much narrower tracks.

current through the head must be set to suit the coercivity of the tape, and is arranged almost to saturate the track. The amplitude of the current is constant, and recording is performed by reversing the direction of the current with respect to time. As the track passes the head, this is converted to the reversal of the magnetic field left on the tape with respect to distance. The magnetic recording is therefore bipolar. Figure 4.4 shows that the recording is actually made just after the trailing pole of the record head where the flux strength from the gap is falling. As in analog recorders, the width of the gap is generally made quite large to ensure that the full thickness of the magnetic coating is recorded, although this cannot be done if the same head is intended to replay.

Figure 4.5 shows what happens when a conventional inductive head, i.e. one having a normal winding, is used to replay the bipolar track made by reversing the record current. The head output is proportional to the rate of change of flux and so only occurs at flux reversals. In other words, the replay head differentiates the flux on the track. The polarity of the resultant pulses alternates as the flux changes and changes back. A circuit is necessary which locates the peaks of the pulses and outputs a signal corresponding to the original record current waveform. There are two ways in which this can be done.

The amplitude of the replay signal is of no consequence and often an AGC system is used to keep the replay signal constant in amplitude. What matters is

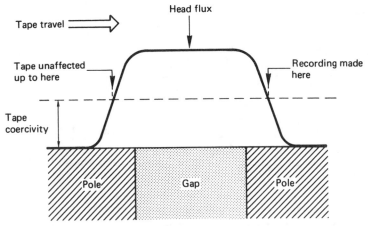

Figure 4.4 The recording is actually made near the trailing pole of the head where the head flux falls below the coercivity of the tape.

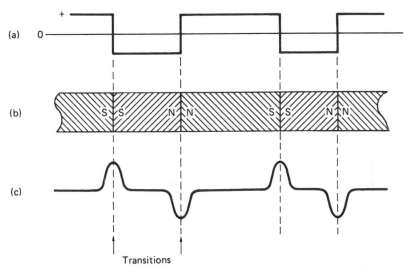

Figure 4.5 Basic digital recording. In (a) the write current in the head is reversed from time to time, leaving a binary magnetization pattern shown in (b). When replayed, the waveform in (c) results because an output is only produced when flux in the head changes. Changes are referred to as transitions.

the time at which the write current, and hence the flux stored on the medium, reverses. This can be determined by locating the peaks of the replay impulses, which can conveniently be done by differentiating the signal and looking for zero crossings. Figure 4.6 shows that this results in noise between the peaks. This problem is overcome by the gated peak detector, where only zero crossings from a pulse which exceeds the threshold will be counted. The AGC system allows the thresholds to be fixed. As an alternative, the record waveform can also be

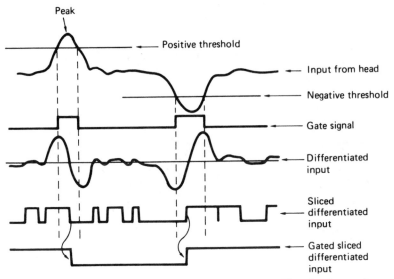

Figure 4.6 Gated peak detection rejects noise by disabling the differentiated output between transitions.

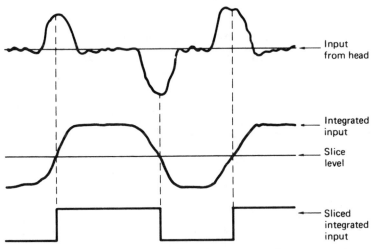

Figure 4.7 Integration method for recreating write-current waveform.

restored by integration, which opposes the differentiation of the head as in Figure 4.7.[1]

The head shown in Figure 4.3 has a frequency response shown in Figure 4.8. At DC there is no change of flux and no output. As a result inductive heads are at a disadvantage at very low speeds. The output rises with frequency until the rise is halted by the onset of thickness loss. As the frequency rises, the recorded wavelength falls and flux from the shorter magnetic patterns cannot be picked up

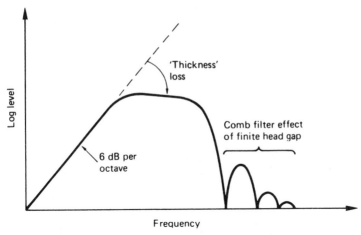

Figure 4.8 The major mechanism defining magnetic channel bandwidth.

so far away. At some point, the wavelength becomes so short that flux from the back of the tape coating cannot reach the head and a decreasing thickness of tape contributes to the replay signal.[2] In digital recorders using short wavelengths to obtain high density, there is no point in using thick coatings. As wavelength further reduces, the familiar gap loss occurs, where the head gap is too big to resolve detail on the track. The construction of the head results in the same action as that of a two-point transversal filter, as the two poles of the head see the tape with a small delay interposed due to the finite gap. As expected, the head response is like a comb filter with the well-known nulls where flux cancellation takes place across the gap. Clearly the smaller the gap the shorter the wavelength of the first null. This contradicts the requirement of the record head to have a large gap. In quality analog audio recorders, it is the norm to have different record and replay heads for this reason, and the same will be true in digital machines which have separate record and playback heads. Clearly where the same pair of heads are used for record and play, the head gap size will be determined by the playback requirement.

As can be seen, the frequency response is far from ideal, and steps must be taken to ensure that recorded data waveforms do not contain frequencies which suffer excessive losses.

A more recent development is the magneto-resistive (M-R) head. This is a head which measures the flux on the tape rather than using it to generate a signal directly. Flux measurement works down to DC and so offers advantages at low tape speeds. Unfortunately flux-measuring heads are not polarity conscious but sense the modulus of the flux and if used directly they respond to positive and negative flux equally, as shown in Figure 4.9. This is overcome by using a small extra winding in the head carrying a constant current. This creates a steady bias field which adds to the flux from the tape. The flux seen by the head is now unipolar and changes between two levels and a more useful output waveform results.

Recorders which have low head-to-medium speed use M-R heads, whereas recorders with high speeds, such as DVTRs, use inductive heads.

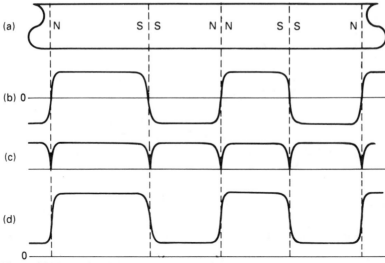

Figure 4.9 The sensing element in a magneto-resistive head is not sensitive to the polarity of the flux, only the magnitude. In (a) the track magnetization is shown and this causes a bidirectional flux variation in the head as in (b), resulting in the magnitude output in (c). However, if the flux in the head due to the track is biased by an additional field, it can be made unipolar as in (d) and the correct waveform is obtained.

Heads designed for use with tape work in actual contact with the magnetic coating. The tape is tensioned to pull it against the head. There will be a wear mechanism and need for periodic cleaning.

In the hard disk, the rotational speed is high in order to reduce access time, and the drive must be capable of staying on line for extended periods. In this case the heads do not contact the disk surface, but are supported on a boundary layer of

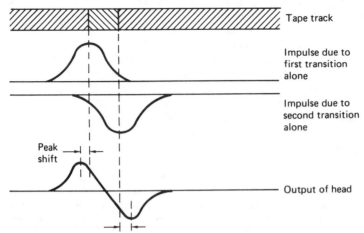

Figure 4.10 Readout pulses from two closely recorded transitions are summed in the head and the effect is that the peaks of the waveform are moved outwards. This is known as peak-shift distortion and equalization is necessary to reduce the effect.

air. The presence of the air film causes spacing loss, which restricts the wavelengths at which the head can replay. This is the penalty of rapid access.

Digital video recorders must operate at high density in order to offer a reasonable playing time. This implies that shortest possible wavelengths will be used. Figure 4.10 shows that when two flux changes, or transitions, are recorded close together, they affect each other on replay. The amplitude of the composite signal is reduced, and the position of the peaks is pushed outwards. This is known as intersymbol interference, or peak-shift distortion, and it occurs in all magnetic media.

The effect is primarily due to high-frequency loss and it can be reduced by equalization on replay, as is done in most tapes, or by pre-compensation on record as is done in hard disks.

4.5 Azimuth recording and rotary heads

Figure 4.11(a) shows that in azimuth recording, the transitions are laid down at an angle to the track by using a head which is tilted. Machines using azimuth recording must always have an even number of heads, so that adjacent tracks can be recorded with opposite azimuth angle. The two track types are usually referred to as A and B. Figure 4.11(b) shows the effect of playing a track with the wrong type of head. The playback process suffers from an enormous azimuth error. The effect of azimuth error can be understood by imagining the tape track to be made from many identical parallel strips. In the presence of azimuth error, the strips at one edge of the track are played back with a phase shift relative to strips at the other side. At some wavelengths, the phase shift will be 180°, and there will be no output; at other wavelengths, especially long wavelengths, some output will reappear. The effect is rather like that of a comb filter, and serves to attenuate crosstalk due to adjacent tracks so that no guard bands are required. Since no tape is wasted between the tracks, more efficient use is made of the tape. The term guard-band-less recording is often used instead of, or in addition to, the term

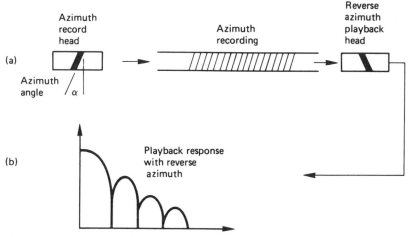

Figure 4.11 In azimuth recording (a), the head gap is tilted. If the track is played with the same head, playback is normal, but the response of the reverse azimuth head is attenuated (b).

azimuth recording. The failure of the azimuth effect at long wavelengths is a characteristic of azimuth recording, and it is necessary to ensure that the spectrum of the signal to be recorded has a small low-frequency content. The signal will need to pass through a rotary transformer to reach the heads, and cannot therefore contain a DC component.

In some rotary head DVTRs there is no separate erase process, and erasure is achieved by overwriting with a new waveform. Overwriting is only successful when there are no long wavelengths in the earlier recording, since these penetrate deeper into the tape, and the short wavelengths in a new recording will not be able to erase them. In this case the ratio between the shortest and longest wavelengths recorded on tape should be limited.

Restricting the spectrum of the code to allow erasure by overwrite also eases the design of the rotary transformer.

4.6 Optical and magneto-optical disks

Optical recorders have the advantage that light can be focused at a distance whereas magnetism cannot. This means that there need be no physical contact between the pickup and the medium and no wear mechanism.

In the same way that the recorded wavelength of a magnetic recording is limited by the gap in the replay head, the density of optical recording is limited by the size of light spot which can be focused on the medium. This is controlled by the wavelength of the light used and by the aperture of the lens. When the light spot is as small as these limits allow, it is said to be diffraction limited.

Some disks can be recorded once, but not subsequently erased or re-recorded. These are known as WORM (write once read many) disks. One type of WORM disk uses a thin metal layer which has holes punched in it on recording by heat from a laser. Others rely on the heat raising blisters in a thin metallic layer by decomposing the plastic material beneath. Yet another alternative is a layer of photochemical dye which darkens when struck by the high-powered recording beam. Whatever the recording principle, light from the pickup is reflected more

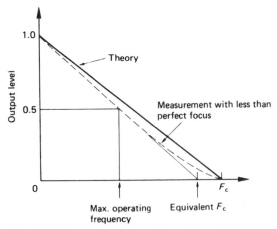

Figure 4.12 Frequency response of laser pickup. Maximum operating frequency is about half of cut-off frequency F_c.

or less, or absorbed more or less, so that the pickup senses a change in reflectivity.

The frequency response of an optical disk is shown in Figure 4.12. The response is best at DC and falls steadily to the optical cut-off frequency. Although the optics work down to DC, this cannot be used for the data recording. DC and low frequencies in the data would interfere with the focus and tracking servos and, as will be seen, difficulties arise when attempting to demodulate a unipolar signal. In practice the signal from the pickup is split by a filter. Low frequencies go to the servos, and higher frequencies go to the data circuitry. As a result the optical disk channel has the same inability to handle DC as does a magnetic recorder, and the same techniques are needed to overcome it.

When a magnetic material is heated above its Curie temperature, it becomes demagnetized, and on cooling will assume the magnetization of an applied field which would be too weak to influence it normally. This is the principle of magneto-optical recording. The heat is supplied by a finely focused laser, and the field is supplied by a coil which is much larger.

Figure 4.13 shows that the medium is initially magnetized in one direction only. In order to record, the coil is energized with a current in the opposite direction. This is too weak to influence the medium in its normal state, but when it is heated by the recording laser beam the heated area will take on the magnetism from the coil when it cools. Thus a magnetic recording with very small dimensions can be made even though the magnetic circuit involved is quite large in comparison.

Readout is obtained using the Kerr effect or the Faraday effect, which are phenomena whereby the plane of polarization of light can be rotated by a magnetic field. The angle of rotation is very small and needs a sensitive pickup. The pickup contains a polarizing filter before the sensor. Changes in polarization

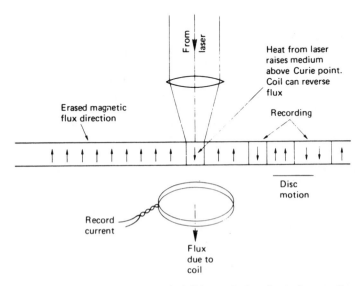

Figure 4.13 The thermomagneto-optical disk uses the heat from a laser to allow magnetic field to record on the disc.

change the ability of the light to get through the polarizing filter and result in an intensity change which once more produces a unipolar output.

The magneto-optic recording can be erased by reversing the current in the coil and operating the laser continuously as it passes along the track. A new recording can then be made on the erased track.

A disadvantage of magneto-optical recording is that all materials having a Curie point low enough to be useful are highly corrodible by air and need to be kept under an effectively sealed protective layer.

The magneto-optical channel has the same frequency response as that shown in Figure 4.12.

4.7 Equalization and data separation

The characteristics of most channels are that signal loss occurs which increases with frequency. This has the effect of slowing down rise times and thereby sloping off edges. If a signal with sloping edges is sliced, the time at which the waveform crosses the slicing level will be changed, and this causes jitter. Figure 4.14 shows that slicing a sloping waveform in the presence of baseline wander causes more jitter.

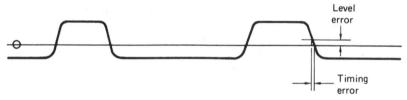

Figure 4.14 A DC offset can cause timing errors.

On a long cable, high-frequency roll-off can cause sufficient jitter to move a transition into an adjacent bit period. This is called intersymbol interference and the effect becomes worse in signals which have greater asymmetry, i.e. short pulses alternating with long ones. The effect can be reduced by the application of equalization, which is typically a high-frequency boost, and by choosing a channel code which has restricted asymmetry.

Compensation for peak shift distortion in recording requires equalization of the channel,[3] and this can be done by a network after the replay head, termed an equalizer or pulse sharpener,[4] as in Figure 4.15(a). This technique uses transversal filtering to oppose the inherent transversal effect of the head. As an alternative, precompensation in the record stage can be used as shown in Figure 4.15(b). Transitions are written in such a way that the anticipated peak shift will move the readout peaks to the desired timing.

The important step of information recovery at the receiver or replay circuit is known as data separation. The data separator is rather like an analog-to-digital converter because the two processes of sampling and quantizing are both present. In the time domain, the sampling clock is derived from the clock content of the channel waveform. In the voltage domain, the process of *slicing* converts the analog waveform from the channel back into a binary representation. The slicer

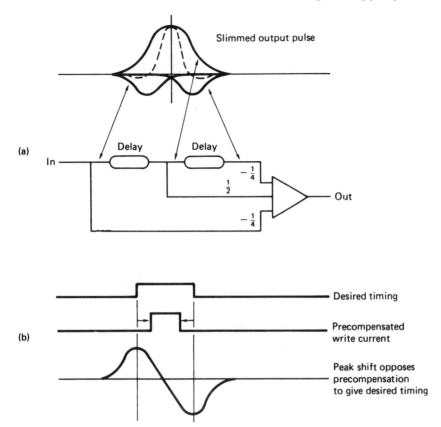

Figure 4.15 Peak-shift distortion is due to the finite width of replay pulses. The effect can be reduced by the pulse slimmer shown in (a) which is basically a transversal filter. The use of a linear operational amplifier emphasizes the analog nature of channels. Instead of replay pulse slimming, transitions can be written with a displacement equal and opposite to the anticipated peak shift as shown in (b).

is thus a form of quantizer which has only 1 bit resolution. The slicing process makes a discrete decision about the voltage of the incoming signal in order to reject noise. The sampler makes discrete decisions along the time axis in order to reject jitter. These two processes will be described in detail.

4.8 Slicing and jitter rejection

The slicer is implemented with a comparator which has analog inputs but a binary output. In a cable receiver, the input waveform can be sliced directly. In an inductive magnetic replay system, the replay waveform is differentiated and must first pass through a peak detector (Figure 4.6) or an integrator (Figure 4.7). The signal voltage is compared with the midway voltage, known as the threshold, baseline or slicing level by the comparator. If the signal voltage is above the threshold, the comparator outputs a high level; if below, a low level results.

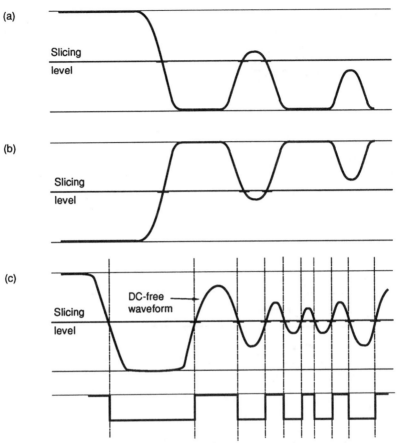

Figure 4.16 Slicing a signal which has suffered losses works well if the duty cycle is even. If the duty cycle is uneven, as in (a), timing errors will become worse until slicing fails. With the opposite duty cycle, the slicing fails in the opposite direction as in (b). If, however, the signal is DC free, correct slicing can continue even in the presence of serious losses, as (c) shows.

Figure 4.16 shows some waveforms associated with a slicer. In (a) the transmitted waveform has an uneven duty cycle. The DC component, or average level, of the signal is received with high amplitude, but the pulse amplitude falls as the pulse gets shorter. Eventually the waveform cannot be sliced.

In (b) the opposite duty cycle is shown. The signal level drifts to the opposite polarity and once more slicing is impossible. The phenomenon is called baseline wander and will be observed with any signal whose average voltage is not the same as the slicing level.

In (c) it will be seen that if the transmitted waveform has a relatively constant average voltage, slicing remains possible up to high frequencies even in the presence of serious amplitude loss, because the received waveform remains symmetrical about the baseline.

It is clearly not possible simply to serialize data in a shift register for so-called direct transmission, because successful slicing can only be obtained if the number of ones is equal to the number of zeros; there is little chance of this happening consistently with real data. Instead, a modulation code or channel code is necessary. This converts the data into a waveform which is DC free or nearly so for the purpose of transmission.

The slicing threshold level is naturally zero in a bipolar system such as magnetic inductive replay or a cable. When the amplitude falls it does so symmetrically and slicing continues. The same is not true of M-R heads and optical pickups, which both respond to intensity and therefore produce a unipolar output. If the replay signal is sliced directly, the threshold cannot be zero, but must be some level approximately half the amplitude of the signal as shown in Figure 4.17(a). Unfortunately when the signal level falls it falls towards zero and not towards the slicing level. The threshold will no longer be appropriate for the signal as can be seen in (b). This can be overcome by using a DC-free coded waveform. If a series capacitor is connected to the unipolar signal from an optical

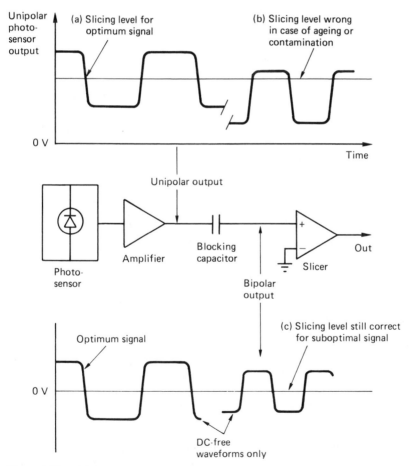

Figure 4.17 (a) Slicing a unipolar signal requires a non-zero threshold. (b) If the signal amplitude changes, the threshold will then be incorrect. (c) If a DC-free code is used, a unipolar waveform can be converted to a bipolar waveform using a series capacitor. A zero threshold can be used and slicing continues with amplitude variations.

pickup, the waveform is rendered bipolar because the capacitor blocks any DC component in the signal. The DC-free channel waveform passes through unaltered. If an amplitude loss is suffered, Figure 4.17(c) shows that the resultant bipolar signal now reduces in amplitude about the slicing level and slicing can continue.

The binary waveform at the output of the slicer will be a replica of the transmitted waveform, except for the addition of jitter or time uncertainty in the position of the edges due to noise, baseline wander, intersymbol interference and imperfect equalization.

Binary circuits reject noise by using discrete voltage levels which are spaced further apart than the uncertainty due to noise. In a similar manner, digital coding combats time uncertainty by making the time axis discrete using events, known as transitions, spaced apart at integer multiples of some basic time period, called a detent, which is larger than the typical time uncertainty. Figure 4.18 shows how this jitter-rejection mechanism works. All that matters is to identify the detent in which the transition occurred. Exactly where it occurred within the detent is of no consequence.

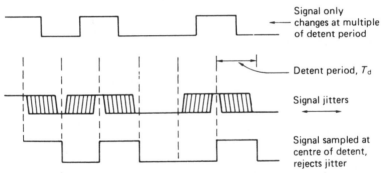

Figure 4.18 A certain amount of jitter can be rejected by changing the signal at multiples of the basic detent period T_d.

As ideal transitions occur at multiples of a basic period, an oscilloscope, which is repeatedly triggered on a channel-coded signal carrying random data, will show an eye pattern if connected to the output of the equalizer. Study of the eye pattern reveals how well the coding used suits the channel. In the case of transmission with a short cable, the losses will be small, and the eye opening will be virtually square except for some edge sloping due to cable capacitance. As cable length increases, the harmonics are lost and the remaining fundamental gives the eyes a diamond shape. The same eye pattern will be obtained with a recording channel where it is uneconomic to provide bandwidth much beyond the fundamental.

Noise closes the eyes in a vertical direction, and jitter closes the eyes in a horizontal direction, as in Figure 4.19. If the eyes remain sensibly open, data separation will be possible. Clearly more jitter can be tolerated if there is less noise, and vice versa. If the equalizer is adjustable, the optimum setting will be where the greatest eye opening is obtained.

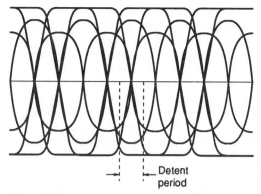

Figure 4.19 A transmitted waveform which is generated according to the principle of Figure 4.18 will appear like this on an oscilloscope as successive parts of the waveform are superimposed on the tube. When the waveform is rounded off by losses, diamond-shaped eyes are left in the centre, spaced apart by the detent period.

In the centre of the eyes, the receiver must make binary decisions at the channel bit rate about the state of the signal, high or low, using the slicer output. As stated, the receiver is sampling the output of the slicer, and it needs to have a sampling clock in order to do that. In order to give the best rejection of noise and jitter, the clock edges which operate the sampler must be in the centre of the eyes.

As has been stated, a separate clock is not practicable in recording or transmission. A fixed-frequency clock at the receiver is of no use as even if it was sufficiently stable, it would not know what phase to run at.

The only way in which the sampling clock can be obtained is to use a phase-locked loop to regenerate it from the clock content of the self-clocking channel-coded waveform. In phase-locked loops, the voltage-controlled oscillator is

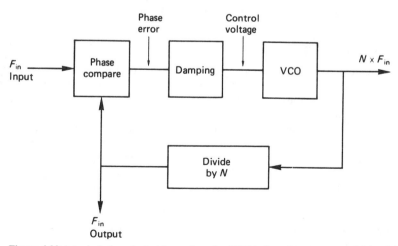

Figure 4.20 A typical phase-locked loop where the VCO is forced to run at a multiple of the input frequency. If the input ceases, the output will continue for a time at the same frequency until it drifts.

driven by a phase error measured between the output and some reference, such that the output eventually has the same frequency as the reference. If a divider is placed between the VCO and the phase comparator, as in Figure 4.20, the VCO frequency can be made to be a multiple of the reference. This also has the effect of making the loop more heavily damped. If a channel-coded waveform is used as a reference to a PLL, the loop will be able to make a phase comparison whenever a transition arrives and will run at the channel bit rate. When there are several detents between transitions, the loop will *flywheel* at the last known frequency and phase until it can rephase at a subsequent transition. Thus a continuous clock is re-created from the clock content of the channel waveform. In a recorder, if the speed of the medium should change, the PLL will change frequency to follow. Once the loop is locked, clock edges will be phased with the average phase of the jittering edges of the input waveform. If, for example, rising edges of the clock are phased to input transitions, then falling edges will be in the centre of the eyes. If these edges are used to clock the sampling process, the maximum jitter and noise can be rejected. The output of the slicer when sampled by the PLL edge at the centre of an eye is the value of a channel bit. Figure 4.21 shows the complete clocking system of a channel code from encoder to data separator. Clearly data cannot be separated if the PLL is not locked, but it cannot be locked until it has seen transitions for a reasonable period. In recorders, which have discontinuous recorded blocks to allow editing, the solution is to precede each data block with a pattern of transitions whose sole purpose is to provide a timing reference for synchronizing the phase-locked loop. This pattern is known as a preamble. In interfaces, the transmission can be continuous and there is no difficulty remaining in lock indefinitely. There will simply be a short delay on first applying the signal before the receiver locks to it.

One potential problem area which is frequently overlooked is to ensure that the VCO in the receiving PLL is correctly centred. If it is not, it will be running with a static phase error and will not sample the received waveform at the centre of the eyes. The sampled bits will be more prone to noise and jitter errors. VCO centring can simply be checked by displaying the control voltage. This should not change significantly when the input is momentarily interrupted.

4.9 Channel coding

In summary, it is not practicable simply to serialize raw data in a shift register for the purpose of recording or for transmission except over relatively short distances. Practical systems require the use of a modulation scheme, known as a channel code, which expresses the data as waveforms which are self-clocking in order to reject jitter, separate the received bits and to avoid skew on separate clock lines. The coded waveforms should further be DC free or nearly so to enable slicing in the presence of losses and have a narrower spectrum than the raw data to make equalization possible.

Jitter causes uncertainty about the time at which a particular event occurred. The frequency response of the channel then places an overall limit on the spacing of events in the channel. Particular emphasis must be placed on the interplay of bandwidth, jitter and noise, which will be shown here to be the key to the design of a successful channel code.

Figure 4.22 shows that a channel coder is necessary prior to the record stage, and that a decoder, known as a data separator, is necessary after the replay stage.

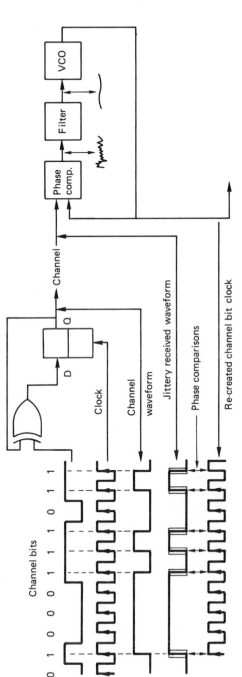

Figure 4.21 The clocking system when channel coding is used. The encoder clock runs at the channel bit rate, and any transitions in the channel must coincide with encoder clock edges. The reason for doing this is that, at the data separator, the PLL can lock to the edges of the channel signal, which represent an intermittent clock, and turn it into a continuous clock. The jitter in the edges of the channel signal causes noise in the phase error of the PLL, but the damping acts as a filter and the PLL runs at the average phase of the channel bits, rejecting the jitter.

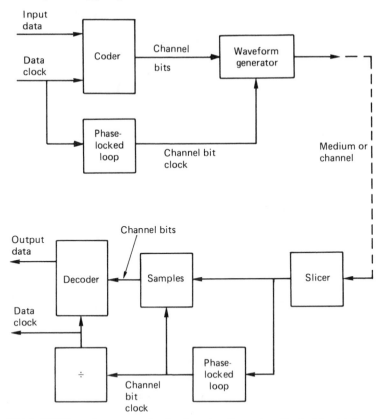

Figure 4.22 The major components of a channel coding system. See text for details.

The output of the channel coder is generally a logic level signal which contains a 'high' state when a transition is to be generated. The waveform generator produces the transitions in a signal whose level and impedance is suitable for driving the medium or channel. The signal may be bipolar or unipolar as appropriate.

Some codes eliminate DC entirely, which is advantageous for cable transmission, optical media and rotary-head recording. Some codes can reduce the channel bandwidth needed by lowering the upper spectral limit. This permits higher linear density, usually at the expense of jitter rejection. Other codes narrow the spectrum by raising the lower limit. A code with a narrow spectrum has a number of advantages. The reduction in asymmetry will reduce peak shift and data separators can lock more readily because the range of frequencies in the code is smaller. In theory the narrower the spectrum the less noise will be suffered, but this is only achieved if filtering is employed. Filters can easily cause phase errors which will nullify any gain.

A convenient definition of a channel code (for there are certainly others) is: 'A method of modulating real data such that they can be reliably received despite the shortcomings of a real channel, while making maximum economic use of the channel capacity'.

The basic time periods of a channel-coded waveform are called positions or detents, in which the transmitted voltage will be reversed or stay the same. The symbol used for the units of channel time is T_d.

One of the fundamental parameters of a channel code is the density ratio (DR). One definition of density ratio is that it is the worst-case ratio of the number of data bits recorded to the number of transitions in the channel. It can also be thought of as the ratio between the Nyquist rate of the data (one-half the bit rate) and the frequency response required in the channel. The storage density of data recorders has steadily increased due to improvements in medium and transducer technology, but modern storage densities are also a function of improvements in channel coding.

As jitter is such an important issue in digital recording and transmission, a parameter has been introduced to quantify the ability of a channel code to reject time instability. This parameter, the jitter margin, also known as the window margin or phase margin (T_w), is defined as the permitted range of time over which a transition can still be received correctly, divided by the data bit-cell period (T).

Since equalization is often difficult in practice, a code which has a large jitter margin will sometimes be used because it resists the effects of intersymbol interference well. Such a code may achieve a better performance in practice than a code with a higher density ratio but poor jitter performance.

A more realistic comparison of code performance will be obtained by taking into account both density ratio and jitter margin. This is the purpose of the figure of merit (FoM), which is defined as DR $\times$ T_w.

4.10 Simple codes

In the non-return-to-zero code shown in Figure 4.23(a), the record current does not cease between bits, but flows at all times in one direction or the other dependent on the state of the bit to be recorded. This results in a replay pulse only when the data bits change from state to another. As a result if one pulse was missed, the subsequent bits would be inverted. This was avoided by adapting the coding such that the record current would change state or invert whenever a data one occurred, leading to the term non-return-to-zero invert or NRZI shown in Figure 4.23(b). In NRZI a replay pulse occurs whenever there is a data one. Clearly neither NRZ or NRZI are self-clocking, but require a separate clock track. Skew between tracks can only be avoided by working at low density and

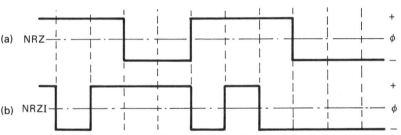

Figure 4.23 In the NRZ code (a) a missing replay pulse inverts every following bit. This was overcome in the NRZI code (b) which reverses write current on a data one.

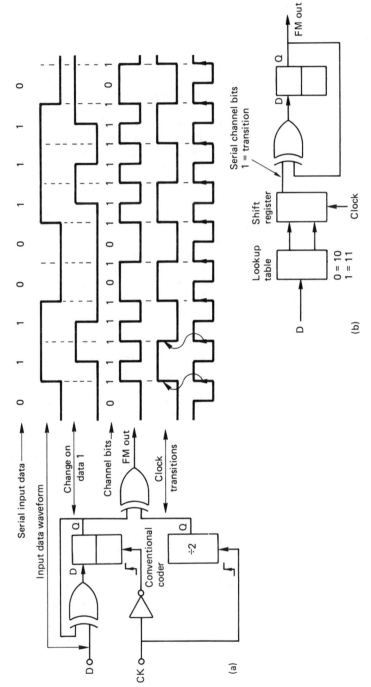

Figure 4.24 In (a) are the FM waveform from a conventional coder and the channel bits which may be used to describe transitions in it. A coder based on a lookup table is shown at (b).

so the system cannot be used directly for digital video. However, virtually all of the codes used for magnetic recording are based on the principle of reversing the record current to produce a transition.

The FM code, also known as Manchester code or bi-phase mark code, shown in Figure 4.24(a), was the first practical self-clocking binary code and it is suitable for both transmission and recording. It is DC free and very easy to encode and decode. It is the code specified for the AES/EBU digital audio interconnect standard. In the field of recording it remains in use today only where density is not of prime importance, for example in SMPTE/EBU timecode for professional audio and video recorders and in floppy disks.

In FM there is always a transition at the bit-cell boundary which acts as a clock. For a data one, there is an additional transition at the bit-cell centre. Figure 4.24(a) shows that each data bit can be represented by two channel bits. For a data zero, they will be 10, and for a data one they will be 11. Since the first bit is always one, it conveys no information, and is responsible for the density ratio of only one-half. Since there can be two transitions for each data bit, the jitter margin can only be half a bit, and the resulting FoM is only 0.25. The high clock content of FM does, however, mean that data recovery is possible over a wide range of speeds; hence the use for timecode. The lowest frequency in FM is due to a stream of zeros and is equal to half the bit rate. The highest frequency is due to a stream of ones, and is equal to the bit rate. Thus the fundamentals of FM are within a band of one octave. Effective equalization is generally possible over such a band. FM is not polarity conscious and can be inverted without changing the data.

Figure 4.24(b) shows how an FM coder works. Data words are loaded into the input shift register which is clocked at the data bit rate. Each data bit is converted to two channel bits in the codebook or lookup table. These channel bits are loaded into the output register. The output register is clocked twice as fast as the input register because there are twice as many channel bits as data bits. The ratio of the two clocks is called the code rate; in this case it is a rate one-half code. Ones in the serial channel bit output represent transitions whereas zeros represent no change. The channel bits are fed to the waveform generator which is a 1 bit delay, clocked at the channel bit rate, and an exclusive OR gate. This changes state when a channel bit one is input. The result is a coded FM waveform where there is always a transition at the beginning of the data bit period, and a second optional transition whose presence indicates a one.

In modified frequency modulation (MFM), also known as Miller code,[5] the highly redundant clock content of FM was reduced by the use of a phase-locked loop in the receiver which could flywheel over missing clock transitions. This technique is implicit in all the more advanced codes. Figure 4.25(a) shows that the bit-cell centre transition on a data one was retained, but the bit-cell boundary transition is now only required between successive zeros. There are still two channel bits for every data bit, but adjacent channel bits will never be one, doubling the minimum time between transitions, and giving a DR of 1. Clearly the coding of the current bit is now influenced by the preceding bit. The maximum number of prior bits which affect the current bit is known as the constraint length L_c, measured in data-bit periods. For MFM $L_c = T$. Another way of considering the constraint length is that it assesses the number of data bits which may be corrupted if the receiver misplaces one transition. If L_c is long, all errors will be burst errors.

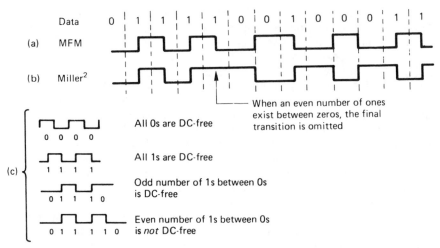

Figure 4.25 MFM or Miller code is generated as shown here. The minimum transition spacing is twice that of FM or PE. MFM is not always DC free as shown in (c). This can be overcome by the modification of (b) which results in the Miller² code.

MFM doubled the density ratio compared with FM and PE without changing the jitter performance; thus the FoM also doubles, becoming 0.5. It was adopted for many rigid disks at the time of its development, and remains in use on double-density floppy disks. It is not, however, DC-free. Figure 4.25(b) shows how MFM can have DC content under certain conditions.

The Miller² code is derived from MFM, and Figure 4.25(c) shows that the DC content is eliminated by a slight increase in complexity.[6,7] Wherever an even number of ones occurs between zeros, the transition at the last one is omitted. This creates two additional, longer run lengths and increases the T_{max} of the code. The decoder can detect these longer run lengths in order to reinsert the suppressed ones. The FoM of Miller² is 0.5, as for MFM. Miller² is used in the D-2 and DCT DVTR formats.

4.11 EFM code in D-3/D-5

Further improvements in coding rely on converting patterns of real data to patterns of channel bits with more desirable characteristics using a conversion table known as a codebook. If a data symbol of m bits is considered, it can have 2^m different combinations. As it is intended to discard undesirable patterns to improve the code, it follows that the number of channel bits n must be greater than m. The number of patterns which can be discarded is:

$$2^n - 2^m$$

One name for the principle is group code recording (GCR), and an important parameter is the code rate, defined as:

$$R = \frac{m}{n}$$

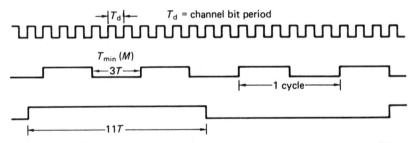

Figure 4.26 A channel code can control its spectrum by placing limits on T_{min} (*M*) and T_{max} which define upper and lower frequencies. The ratio of T_{max}/T_{min} determines the asymmetry of waveform and predicts DC content and peak shift.

It will be evident that the jitter margin T_w is numerically equal to the code rate, and so a code rate near to unity is desirable. The choice of patterns which are used in the codebook will be those which give the desired balance between clock content, bandwidth and DC content.

Figure 4.26 shows that the upper spectral limit can be made to be some fraction of the channel bit rate according to the minimum distance between ones in the channel bits. This is known as T_{min}, also referred to as the minimum transition parameter *M*, and in both cases is measured in data bits *T*. It can be obtained by multiplying the number of channel detent periods between transitions by the code rate.

The 1/2 inch D-3 and D-5 formats use a group code in which $m = 8$ and $n = 14$ so the code rate is 0.57. The code is called 8,14 after the main parameters.

The code used in D-3/D-5 uses the convention in which a channel bit one represents a high in the recorded waveform. In this convention a flux reversal or transition will be written when the channel bits change.

In Figure 4.27 it is shown that the upper spectral limit can be made to be some fraction of the channel bit rate according to the minimum distance between transitions in the channel bits, which in 8,14 is two channel bits. This is known as T_{min}, also referred to as the minimum transition parameter *M*, and in both cases is measured in data bits *T*. It can be obtained by multiplying the number of channel detent periods between transitions by the code rate. In fact T_{min} is numerically equal to the density ratio.

$$T_{min} = M = DR = \frac{2 \times 8}{14} = 1.14$$

This DR is a little better than the figure of 1 for the codes used in D-1 and D-2.

The figure of merit is:

$$FoM = DR \times T_w = \frac{2 \times 8^2}{14^2} = 0.65$$

since:

$$T_w = \frac{m}{n} = \frac{8}{14}$$

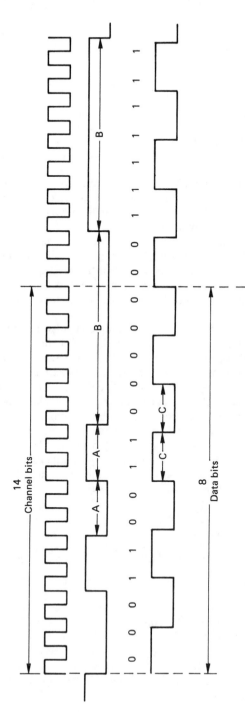

Figure 4.27 In the 8,14 code of D-3, eight data-bit periods are divided into 14 channel-bit periods. However, the shortest run length allowed in the channel is 2 bits, shown at A. This restriction is obtained by selecting 14 bit patterns from those available. The longest run length in the code is seven channel bits. Note that the shortest run length A is 14% longer than the shortest run length in the raw data C. Thus density ratio DR is 1.14. Using this code 14% more data can be recorded with the same wavelength on tape as a simple code.

Figure 4.27 also shows that the lower spectral limit is influenced by the maximum distance between transitions T_{max}, which also determines the minimum clock content. This is also obtained by multiplying the maximum number of detent periods between transitions by the code rate. In 8,14 code this is seven channel bits, and so:

$$T_{max} = \frac{7 \times 8}{14} = 4$$

and the maximum/minimum ratio P is:

$$P = \frac{4}{1.14} = 3.51$$

The length of time between channel transitions is known as the run length. Another name for this class is the run-length-limited (RLL) codes.[8] Since 8 data bits are considered as one symbol, the constraint length L_c will be increased in this code to at least 8 bits.

In practice, the junction of some adjacent channel symbols may violate coding rules, and it is necessary to extend the codebook so that the original data can be represented by a number of alternative codes at least one of which will be acceptable. This is known as substitution. Owing to the coding convention used, which generates a transition when the channel bits change, transitions will also be generated at the junction of two channel symbols if the adjacent bits are different. It will be seen in Figure 4.27 that the presence of a junction transition can be controlled by selectively inverting all of the bits in the second channel symbol. Thus for every channel bit pattern in the code, an inverted version also exists.

Figure 4.28 shows part of the 8,14 code (EFM) used in the D-3 format. As stated, 8 bit data symbols are represented by 14 bit channel symbols. There are 256 combinations of eight data bits, whereas 14 bits have 2^{14} or 16 384 combinations. The initial range of possible codes is reduced by the requirements of the maximum and minimum run-length limits and by a requirement that there shall not be more than 5 identical bits in the first 6 bits of the code and not more than 6 in the last 7 bits as shown in Figure 4.29.

One of the most important parameters of a channel pattern is the code digital sum (CDS) shown in Figure 4.30. This is the number of channel ones minus the number of channel zeros. As the CDS represents the DC content or average voltage of the channel pattern it is to be kept to a minimum.

There are only 118 codes (and 118 inverse codes) which are DC free (CDS = 0) as well as meeting the other constraints. In order to obtain 256 data combinations, codes with non-zero CDS have to be accepted. The actual codes used in 8,14 have CDS of 0, ±2 or ±4.

The CDS is a special case of a more general parameter called the digital sum value (DSV). DSV is a useful way of predicting how an analog channel such as a tape/head system will handle a binary waveform. In a stream of bits, a one causes one to be added to the DSV whereas a zero causes one to be subtracted. Thus DSV is a form of running discrete integration. Figure 4.31(a) shows how the DSV varies along the time axis. The End DSV is the DSV at the end of a string of channel symbols. Clearly CDS is the End DSV of one code pattern in isolation. The next End DSV is the current one plus the CDS of the next symbol as shown in Figure 4.31(b).

Data	Code A (begins with 0)	CDS	Code B (begins with 1)	CDS	Code C (begins with 0)	CDS	Code D (begins with 1)	CDS
00 ↑								
42	01100011000111	0	10011100111000	0	01100011000111	0	10011100111000	0
43	01100001111100	0	10011110000011	0	01100001111100	0	10011110000011	0
44	01100001111001	0	10011110000110	0	01100001111001	0	10011110000110	0
45	01100001110011	0	10011110001100	0	01100001110011	0	10011110001100	0
46	01100001100111	0	10011110011000	0	01100001100111	0	10011110011000	0
47	01100000111110	0	10011111000001	0	01100000111110	0	10011111000001	0
48	01100000111111	0	10011111100000	0	01100000011111	0	10011111100000	0
49	01111100000001	-2	10000000110011	-2	01111111001100	-4	10000011111110	2
50	01111001100000	-2	10000000111001	-2	01111111000110	-4	10000110011111	2
51	01111000010000	-2	10000000111100	-2	01111111000011	-4	10000111001111	2
52	01111000011000	-2	10000001100011	-2	01111110011100	-4	10000111100111	2
53	01111000001100	-2	10000001100110	-2	01111110011001	-4	10000111110011	2
54	01111000000110	-2	10000001110001	-2	01111111000110	-4	10000111111001	2
55	01111000000011	-2	10000001111000	-2	01111110000111	-4	10000111111100	2
56	01110011100000	-2	10000011000011	-2	01111100111100	-4	10001100011111	2
57	01110011000001	-2	10000011000110	-2	01111100111001	-4	10001100111110	2
↓ 255								

Figure 4.28 Part of the codebook for D-3 EFM code. For every 8 bit data word (left) there are four possible channel-bit patterns, two of which begin with 0 and two of which begin with 1. This allows successive symbols to avoid violating the minimum run-length limit (A in Figure 4.27) at the junction of the two symbols. See Figures 4.29 and 4.30 for further details.

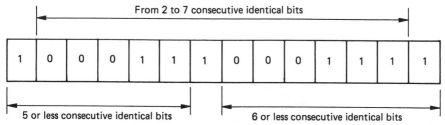

Figure 4.29 The selected 14 bit patterns follow the rules that there cannot be more than five consecutive identical bits at the beginning or more than six at the end. In addition the code digital sum (CDS) cannot exceed 4. See Figure 4.30 for derivation of CDS.

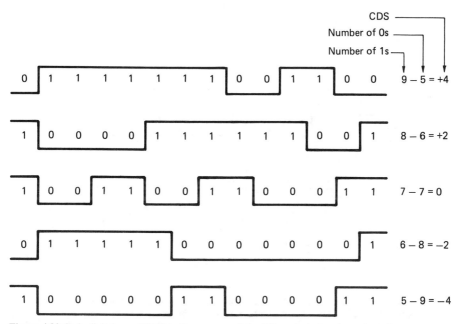

Figure 4.30 Code digital sum (CDS) is the measure of the DC content of a given channel symbol. It is obtained by subtracting the number of zeros in the symbol from the number of ones. Shown above are five actual 14 bit symbols from the D-3 code showing the range of CDS allowed from −4 to +4. Although CDS of zero is optimum, there are not enough patterns with zero CDS to represent all combinations of eight data bits.

The Absolute DSV of a symbol is shown in Figure 4.31(c) This is obtained by finding the peak DSV. Large Absolute DSVs are associated with low frequencies and low clock content.

When encoding is performed, each 8 bit data symbol selects four locations in the lookup table, each of which contains a 14 bit pattern. The recording is made by selecting the most appropriate one of four candidate channel bit patterns. The decoding process is such that any of the four channel patterns will decode to the same data.

The four possible channel symbols for each data byte are classified according to Figure 4.28 into types A, B, C or D. Since 8,14 coding requires alternative

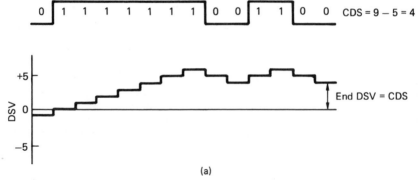

Figure 4.31(a) DSV is obtained by subtracting one for every zero, and adding one for every one which passes during the integration time. Thus DSV changes every channel bit. Note that end DSV = CDS.

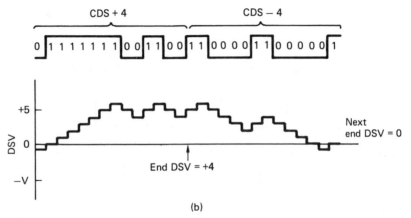

Figure 4.31(b) The next end DSV is the current end DSV plus the CDS of the next symbol. Note how the choice of a CDS −4 symbol after a +4 symbol brings DSV back to zero.

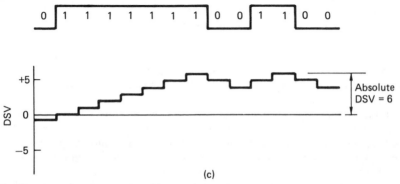

Figure 4.31(c) The absolute or peak DSV is the greatest value the DSV can have during the symbol. As shown here, it can be greater than CDS.

+2 end DSV end pattern of channel bits	−2 end DSV end pattern of channel bits	Priority
. . . xxxxx110	. . . xxxxx001	4
. . . xxxx1100	. . . xxxx0011	1
. . . xxx11000	. . . xxx00111	2
. . . xx110000	. . . xx001111	3
. . . x1100000	. . . x0011111	8
. . . xxxxx001	. . . xxxxx110	10
. . . xxxx0011	. . . xxxx1100	5
. . . xxx00111	. . . xxx11000	6
. . . xx001111	. . . xx110000	7
. . . x0011111	. . . x1100000	9
. . . 00111111	. . . 11000000	11

x: Don't care bit

Figure 4.32 Codes having end DSV of ±2 are prioritized according to the above table as part of the selection process.

inverse symbols, two of the candidates are simply bit inversions of the other two. Where codes are not DC free there may be a pair of +2 CDS candidates and their −2 CDS inverses or a +2 CDS and a +4 CDS candidate and their inverses which will of course have −2 and −4 CDS. In the case of all but two of the DC-free codes the two candidates are one and the same code and the four lookup table locations contain two identical codes and two identical inverse codes. As a result of some of the codes being identical, although there are 1024 locations in the table, only 770 different 14 bit patterns are used, representing 4.7% of the total.

Code classes are further subdivided into class numbers, which go from 1 to 5 and specify the number of identical channel bits at the beginning of the symbol and a Priority Number which only applies to codes of CDS ±2 and is obtained from the channel bit pattern at the end of the code according to Figure 4.32.

Clearly the junction of two channel patterns cannot be allowed to violate the run-length limits. As the next code cannot have more than five consecutive identical bits at the beginning, the previous code could end with up to two bits in the same state without exceeding the maximum run length of seven. If this limit were to be exceeded by one of the four candidate codes, it would be rejected and one of those having an earlier transition would be chosen instead. Similarly if the previous code ends in 01 or 10, the first bit of the next code must be the same as the last bit of the previous code or the minimum run-length limit will be violated. The run-length limits can always be met because every code has an inverse so out of the four possible channel symbols available for a given data byte two of them will begin with 0 and two begin with 1. In some cases, such as where the first code ends in 1100, up to four of the candidates for the next code could meet the run-length limits in some cases. In this case the best candidate will be chosen to optimize some other parameter.

In order to follow how the encoder selects the best channel pattern it is necessary to discuss the criteria that are used. There are a number of these, some of which are compulsory, such as the run-length limits, and others which will be met if possible on a decreasing scale of importance. If a higher criterion cannot be met the decision will still attempt to meet as many of the lower ones as possible.

The overall goal is to meet the run-length limits with a sequence of symbols which has the highest clock content, lowest LF and DC content and the least asymmetry to reduce peak shift. Some channel symbols will be better than others, a phenomenon known as pattern sensitivity. The least optimal patterns are not so bad that they will *cause* errors, but they will be more prone to errors due to other causes. Minimizing the use of sensitive waveforms will enhance the data reliability and is as good as an improvement in the signal-to-noise ratio.

As there are 2^{10} different patterns, there will be 2^{20} different combinations of two patterns. Clearly it is out of the question to create a lookup table to determine how best to merge two patterns. It has to be done algorithmically.

4.12 Randomizing

Randomising is not a channel code, but a technique which can be used in conjunction with a channel code. Randomizing with NRZI (RNRZI) is used in the D-1 format and in conjunction with partial response (see Section 4.13) in Digital Betacam. It is also used in conjunction with the 8,14 code of D-3/D-5 and in NICAM 728. Figure 4.33 shows that the randomizing system is arranged outside the channel coder.

NRZ has a DR of 1 and a jitter window of 1 and so has a FoM of 1 which is better than the group codes. It does, however, suffer from an unconstrained spectrum and poor clock content. This can be overcome using randomizing. Figure 4.34 shows that, at the encoder, a pseudo-random sequence is added

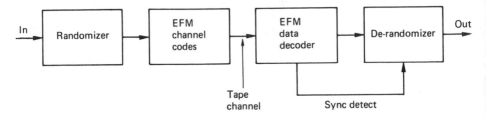

Figure 4.33 The randomizing system of D-3 is arranged outside the EFM channel coder. The pseudo-random sequence added during recording must be generated in a synchronized manner during replay so it can be subtracted. This is done by the EFM sync detector. Note that the sync pattern is not randomized!

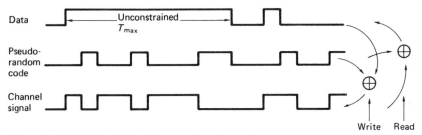

Figure 4.34 Modulo-2 addition with a pseudo-random code removes unconstrained runs in real data. Identical process must be provided on replay.

modulo-2 to the serial data and the resulting ones generate transitions in the channel. This process drastically reduces T_{max} and reduces DC content. At the receiver the transitions are converted back to a serial bit stream to which the same pseudo-random sequence is again added modulo-2. As a result the random signal cancels itself out to leave only the serial data, provided that the two pseudo-random sequences are synchronized to bit accuracy.

The 8,14 code displays pattern sensitivity because some waveforms are more sensitive to peak shift distortion than others. Pattern sensitivity is only a problem if a sustained series of sensitive symbols needs to be recorded. Randomizing ensures that this cannot happen because it breaks up any regularity or repetition in the data. The data randomizing is performed by using the exclusive OR function of the data and a pseudo-random sequence as the input to the channel coder. On replay the same sequence is generated, synchronized to bit accuracy, and the exclusive OR of the replay bit stream and the sequence is the original data.

The randomizing polynomial and one way in which it can be implemented are shown in Figure 4.35. As the polynomial generates a maximum length sequence from an 8 bit wordlength, the sequence length is given by $2^8 - 1 = 255$. The sequence would repeat endlessly but for the fact that it is preset to 80_{16} at the beginning of each sync block immediately after the sync pattern is detected. Figure 4.36 shows the randomizing sequence starting from 80_{16}.

Clearly the sync pattern cannot be randomized, since this causes a Catch-22 situation where it is not possible to synchronize the sequence for replay until the sync pattern is read, but it is not possible to read the sync pattern until the sequence is synchronized!

The randomizing in D-3 is clearly block based, since this matches the block structure on tape. Where there is no obvious block structure, convolutional or endless randomizing can be used. This is the approach used in the scrambled serial digital video interconnect (SDI) which allows composite or component video of up to 10 bit wordlength to be sent serially.

In convolutional randomizing, the signal sent down the channel is the serial data waveform which has been convolved with the impulse response of a digital filter. On reception the signal is deconvolved to restore the original data. Figure 4.37(a) shows that the filter is an infinite-impulse response (IIR) filter which has recursive paths from the output back to the input. As it is a 1 bit filter its output cannot decay, and once excited, it runs indefinitely. The filter is followed by a transition generator which consists of a 1 bit delay and an exclusive OR gate. An

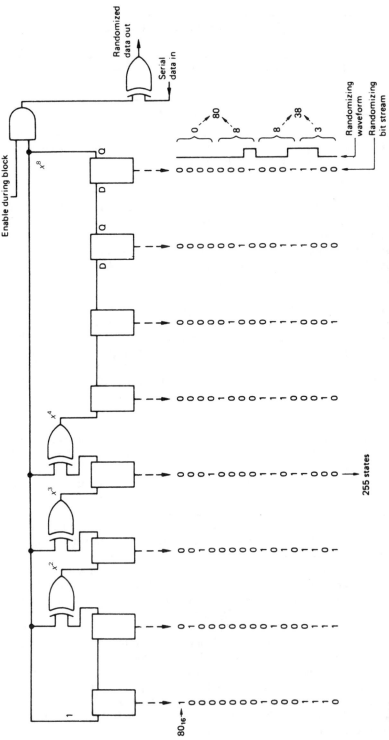

Figure 4.35 The polynomial generator circuit shown here calculates $x^8 + x^4 + x^3 + x^2 + 1$ and is preset to 80_{16} at the beginning of every sync block. When the generator is clocked it will produce a Galois field having 255 states. The right-hand bit of each field element becomes the randomizing bit stream and is fed to an exclusive OR gate in the data stream. Randomizing is disabled during preambles and sync patterns. It is also possible to randomize using a counter working at byte rate which addresses a PROM. Eight exclusive OR gates will then be needed to randomize 1 byte at a time.

80	38	D2	81	49	76	82	DA	9A	86	6F	AF	8B	B0	F1	9C
D1	12	A5	72	37	EF	97	59	31	B8	EA	53	C8	3F	F4	58
40	1C	E9	C0	24	38	41	6D	4D	C3	B7	D7	45	D8	78	CE
68	89	52	B9	9B	F7	CB	AC	18	5C	F5	29	E4	1F	7A	2C
20	8E	74	60	92	9D	A0	B6	A6	E1	DB	EB	22	6C	3C	67
B4	44	A9	DC	CD	FB	65	56	0C	AE	FA	14	F2	0F	3D	16
10	47	3A	30	C9	4E	50	5B	D3	F0	ED	75	11	36	9E	33
5A	A2	54	EE	E6	FD	32	2B	06	57	7D	0A	F9	87	1E	0B
88	23	1D	98	64	27	A8	AD	69	F8	F6	BA	08	1B	CF	19
2D	51	2A	77	F3	7E	99	15	83	AB	3E	85	FC	43	8F	05
C4	91	0E	4C	B2	13	D4	D6	34	7C	7B	5D	84	8D	E7	8C
96	28	95	BB	79	BF	CC	8A	C1	55	9F	42	FE	A1	C7	02
E2	48	07	26	D9	09	6A	6B	1A	BE	BD	2E	C2	C6	73	46
4B	94	CA	DD	BC	5F	66	C5	E0	AA	4F	21	FF	D0	63	01
71	A4	03	93	EC	04	B5	35	0D	DF	5E	17	61	E3	39	A3
25	4A	E5	6E	DE	2F	B3	62	70	D5	A7	90	7F	E8	B1	

Figure 4.36 The sequence which results when the randomizer of Figure 4.35 is allowed to run.

input 1 results in an output transition on the next clock edge. An input 0 results in no transition.

A result of the infinite-impulse response of the filter is that frequent transitions are generated in the channel which result in sufficient clock content for the phase-locked loop in the receiver.

Transitions are converted back to ones by a differentiator in the receiver. This consists of a 1 bit delay with an exclusive OR gate comparing the input and the output. When a transition passes through the delay, the input and the output will be different and the gate outputs a 1 which enters the deconvolution circuit.

Figure 4.37(b) shows that in the deconvolution circuit a data bit is simply the exclusive OR of a number of channel bits at a fixed spacing. The deconvolution is implemented with a shift register having the exclusive OR gates connected in a reverse pattern to that in the encoder. The same effect as block randomizing is obtained, in that long runs are broken up and the DC content is reduced, but it has the advantage over block randomizing that no synchronizing is required to remove the randomizing, although it will still be necessary for deserialization. Clearly the system will take a few clock periods to produce valid data after commencement of transmission; this is no problem on a permanent-wired connection where the transmission is continuous.

4.13 Partial response

It has been stated that a magnetic head acts as a transversal filter, because it has two poles. In addition the output is differentiated, so that the head may be thought of as a $(1-D)$ impulse response system, where D is the delay which is a function of the tape speed and gap size. It is this delay which results in intersymbol interference. Conventional equalizers attempt to oppose this effect, and succeed in raising the noise level in the process of making the frequency response linear. Figure 4.38 shows that the frequency response necessary to pass data with insignificant peak shift is a bandwidth of half the bit rate, which is the Nyquist rate. In Class IV partial response, the frequency response of the system is made to have nulls at DC and at the bit rate. Such a frequency response is particularly advantageous for rotary-head recorders as it is DC free and the low-frequency content is minimal, hence the use in Digital Betacam. The required response is

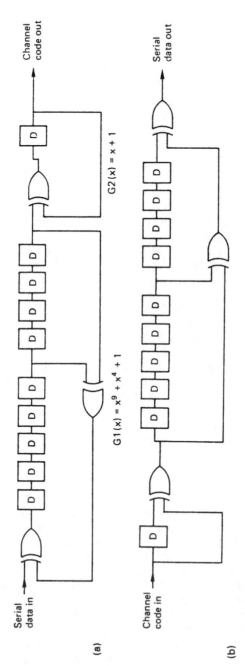

Figure 4.37 (a) Convolutional randomizing encoder transmits exclusive OR of 3 bits at a fixed spacing in the data. One bit delay, far right, produces channel transitions from data ones. Decoder, (b), has opposing 1 bit delay to return from transitions to data levels, followed by an opposing shift register which exactly reverses the coding process.

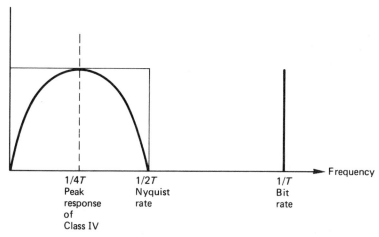

1/4T
Peak
response
of
Class IV

1/2T
Nyquist
rate

1/T
Bit
rate

Frequency

Figure 4.38 Class IV response has spectral nulls at DC and the Nyquist rate, giving a noise advantage, since magnetic replay signal is weak at both frequencies in a high-density channel.

achieved by an overall impulse response of $(1 - D^2)$ where D is now the bit period. There are a number of ways in which this can be done.

If the head gap is made equal to one bit, the $(1 - D)$ head response may be converted to the desired response by the use of a $(1 + D)$ filter, as in Figure 4.39(a).[9] Alternatively, a head of unspecified gapwidth may be connected to an integrator, and equalized flat to reproduce the record current waveform before being fed to a $(1 - D^2)$ filter as in Figure 4.39(b).[10]

The result of both of these techniques is a ternary signal. The eye pattern has two sets of eyes as in Figure 4.39(c).[11] When slicing such a signal, a smaller amount of noise will cause an error than in the binary case.

The treatment of the signal thus far represents an equalization technique, and not a channel code. However, to take full advantage of Class IV partial response, suitable precoding is necessary prior to recording, which does then constitute a channel-coding technique. This precoding is shown in Figure 4.40(a). Data are added modulo-2 to themselves with a 2 bit delay. The effect of this precoding is that the outer levels of the ternary signals, which represent data ones, alternate in polarity on all odd bits and on all even bits. This is because the precoder acts like two interleaved 1 bit delay circuits, as in Figure 4.40(b). As this alternation of polarity is a form of redundancy, it can be used to recover the 3 dB SNR loss encountered in slicing a ternary eye pattern. Viterbi decoding[12] can be used for this purpose. In Viterbi decoding, each channel bit is not sliced individually; the slicing decision is made in the context of adjacent decisions. Figure 4.41 shows a replay waveform which is so noisy that, at the decision point, the signal voltage crosses the centre of the eye, and the slicer alone cannot tell whether the correct decision is an inner or an outer level. In this case, the decoder essentially allows both decisions to stand, in order to see what happens. A symbol representing indecision is output. It will be seen from the figure that as subsequent bits are received, one of these decisions will result in an absurd situation, which indicates that the other decision was the right one. The decoder can then locate the undecided symbol and set it to the correct value.

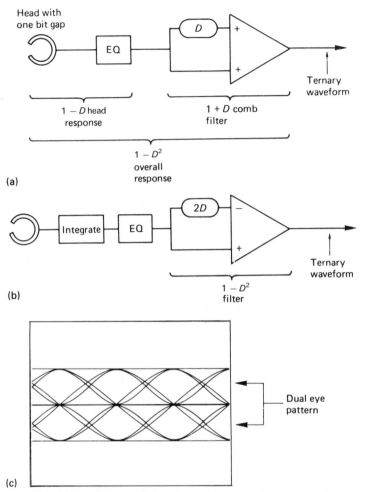

Figure 4.39 (a), (b) Two ways of obtaining partial response. (c) Characteristic eye pattern of ternary signal.

Viterbi decoding requires more information about the signal voltage than a simple binary slicer can discern. Figure 4.42 shows that the replay waveform is sampled and quantized so that it can be processed in digital logic. The sampling rate is obtained from the embedded clock content of the replay waveform. The digital Viterbi processing logic must be able to operate at high speed to handle serial signals from a DVTR head. Its application in Digital Betacam is eased somewhat by the adoption of data reduction which reduces the data rate at the heads by a factor of two.

Clearly a ternary signal having a dual eye pattern is more sensitive than a binary signal, and it is important to keep the maximum run length T_{max} small in order to have accurate AGC. The use of pseudo-random coding along with partial

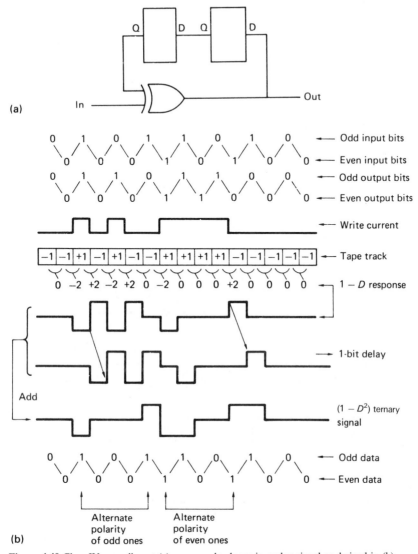

Figure 4.40 Class IV precoding at (a) causes redundancy in replay signal as derived in (b).

response equalization and precoding is a logical combination.[13] There is then considerable overlap between the channel code and the error-correction system. Viterbi decoding is primarily applicable to channels with random errors due to Gaussian statistics, and they cannot cope with burst errors. In a head-noise-limited system, however, the use of a Viterbi detector could increase the power of a separate burst-error-correction system by relieving it of the need to correct random errors due to noise. The error-correction system could then concentrate on correcting burst errors unimpaired.

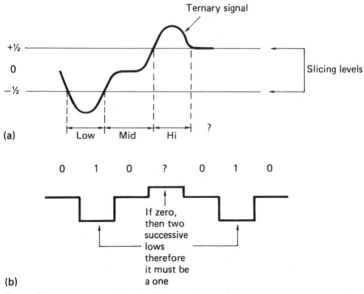

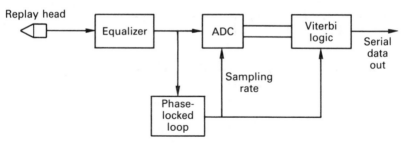

Figure 4.41 (a) A ternary signal suffers a noise penalty because there are two slicing levels. (b) The redundancy is used to determine the bit value in the presence of noise. Here the pulse height has been reduced to make it ambiguous 1/0, but only 1 is valid as zero violates the redundancy rules.

Figure 4.42 A Viterbi decoder is implemented in the digital domain by sampling the replay waveform with a clock locked to the embedded clock of the channel code.

4.14 Synchronizing

Once the PLL in the data separator has locked to the clock content of the transmission, a serial channel bit stream and a channel bit clock will emerge from the sampler. In a group code, it is essential to know where a group of channel bits begins in order to assemble groups for decoding to data bit groups. In a randomizing system it is equally vital to know at what point in the serial data stream the words or samples commence. In serial transmission and in recording, channel bit groups or randomized data words are sent one after the other, one bit at a time, with no spaces in between, so that although the designer knows that a data block contains, say, 128 bytes, the receiver simply finds 1024 bits in a row. If the exact position of the first bit is not known, then it is not possible to put all the bits in the right places in the right bytes, a process known as deserializing. The effect of sync slippage is devastating, because a 1 bit disparity between the bit count and the bit stream will corrupt every symbol in the block.

The synchronization of the data separator and the synchronization to the block format are two distinct problems which are often solved by the same sync pattern. Deserializing requires a shift register which is fed with serial data and read out once per word. The sync detector is simply a set of logic gates which are arranged to recognize a specific pattern in the register. The sync pattern is either identical for every block or has a restricted number of versions and it will be recognized by the replay circuitry and used to reset the bit count through the block. Then by counting channel bits and dividing by the group size, groups can be deserialized and decoded to data groups. In a randomized system, the pseudo-random sequence generator is also reset. Then by counting derandomized bits from the sync pattern and dividing by the wordlength enables the replay circuitry to deserialize the data words.

Even if a specific code were excluded from the recorded data so it could be used for synchronizing, this cannot ensure that the same pattern cannot be falsely created at the junction between two allowable data words. Figure 4.43 shows

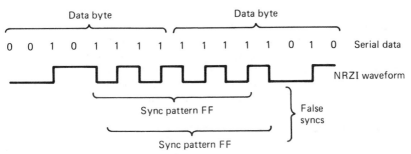

Figure 4.43 Concatenation of two words can result in the accidental generation of a word which is reserved for synchronizing.

how false synchronizing can occur due to concatenation. It is thus not practical to use a bit pattern which is a data code value in a simple synchronizing recognizer. The problem is overcome in NICAM 728 by using the fact that sync patterns occur exactly once per millisecond or 728 bits. The sync pattern of NICAM 728 is just a bit pattern and no steps are taken to prevent it from appearing in the randomized data. If the pattern is seen by the recognizer, the recognizer is disabled for the rest of the frame and only enabled when the next sync pattern is expected. If the same pattern recurs every millisecond, a genuine sync condition exists. If it does not, there was a false sync and the recognizer will be enabled again. As a result it will take a few milliseconds before sync is achieved, but once achieved it should not be lost unless the transmission is interrupted. This is fine for the application and no-one objects to the short mute of the NICAM sound during a channel switch. The principle cannot, however, be used for recording because channel interruptions are more frequent due to head switches and dropouts and loss of several blocks of data due to a single dropout is unacceptable.

In run-length-limited codes this is not a problem. The sync pattern is no longer a data-bit pattern but is a specific waveform. If the sync waveform contains run lengths which violate the normal coding limits, there is no way that these run lengths can occur in encoded data, nor any possibility that they will be interpreted as data. They can, however, be readily detected by the replay circuitry.

In a group code there are many more combinations of channel bits than there are combinations of data bits. Thus after all data-bit patterns have been allocated group patterns, there are still many unused group patterns which cannot occur in the data. With care, group patterns can be found which cannot occur due to the concatenation of any pair of groups representing data. These are then unique and can be used for synchronizing.

4.15 Basic error correction

There are many different types of recording and transmission channels and consequently there will be many different mechanisms which may result in errors. Bit errors in video cause 'sparkles' in the picture whose effect depends upon the significance of the affected bit.

In magnetic recording, data can be corrupted by mechanical problems such as media dropout and poor tracking or head contact, or Gaussian thermal noise in replay circuits and heads. In optical recording, contamination of the medium interrupts the light beam. When group codes are used, a single defect in a group changes the group symbol and may cause errors up to the size of the group. Single-bit errors are therefore less common in group-coded channels. Inside equipment, data are conveyed on short wires and the noise environment is under the designer's control. With suitable design techniques, errors can be made effectively negligible, whereas in communication systems there is considerably less control of the electromagnetic environment.

Irrespective of the cause, all of these mechanisms cause one of two effects. There are large isolated corruptions, called error bursts, where numerous bits are corrupted all together in an area which is otherwise error free, and there are random errors affecting single bits or symbols. Whatever the mechanism, the result will be that the received data will not be exactly the same as those sent. In binary the discrete bits will each be either right or wrong. If a binary digit is known to be wrong, it is only necessary to invert its state and then it must be right. Thus error correction itself is trivial; the hard part is working out *which* bits need correcting.

There are a number of terms which have idiomatic meanings in error correction. The raw BER (bit error rate) is the error rate of the medium, whereas the residual or uncorrected BER is the rate at which the error-correction system fails to detect or miscorrects errors. In practical digital systems, the residual BER is negligibly small. If the error correction is turned off, the two figures become the same.

Error correction works by adding some bits to the data which are calculated from the data. This creates an entity called a codeword which spans a greater length of time than one bit alone. The statistics of noise means that whilst one bit may be lost in a codeword, the loss of the rest of the codeword because of noise is highly improbable. As will be described later in this chapter, codewords are designed to be able to correct totally a finite number of corrupted bits. The greater the timespan over which the coding is performed, or, on a recording medium, the greater area over which the coding is performed, the greater will be the reliability achieved, although this does mean that an encoding delay will be experienced on recording, and a similar or greater decoding delay on reproduction.

Shannon[14] disclosed that a message can be sent to any desired degree of accuracy provided that it is spread over a sufficient timespan. Engineers have to compromise, because an infinite coding delay in the recovery of an error-free signal is not acceptable. Digital interfaces such as SDI do not employ error correction because the build-up of coding delays in large systems is unacceptable.

If error correction is necessary as a practical matter, it is then only a small step to put it to maximum use. All error correction depends on adding bits to the original message, and this of course increases the number of bits to be recorded, although it does not increase the information recorded. It might be imagined that error correction is going to reduce storage capacity, because space has to be found for all the extra bits. Nothing could be further from the truth. Once an error-correction system is used, the signal-to-noise ratio of the channel can be reduced, because the raised BER of the channel will be overcome by the error-correction system. Reduction of the SNR by 3 dB in a magnetic tape track can be achieved by halving the track width, provided that the system is not dominated by head or preamplifier noise. This doubles the recording density, making the storage of the additional bits needed for error correction a trivial matter. In short, error correction is not a nuisance to be tolerated; it is a vital tool needed to maximize the efficiency of recorders. Digital video or audio recording would not be economically viable without it.

Figure 4.44 shows the broad subdivisions of error handling. The first stage might be called error avoidance and includes such measures as creating bad block files on hard disks or using verified media. Placing the audio blocks near to the centre of the tape is a further example. The data pass through the channel, which causes whatever corruptions it feels like. On receipt of the data the occurrence of errors is first detected, and this process must be extremely reliable, as it does not matter how effective the correction or how good the concealment algorithm if it is not known that they are necessary! The detection of an error then results in a course of action being decided.

In most cases of digital video or audio replay a retry is not possible because the data are required in real time. However, if a disk-based system is transferring to tape for the purpose of backup, real-time operation is not required. If the disk drive detects an error a retry is easy as the disk is turning at several thousand rpm and will quickly re-present the data. An error due to a dust particle may not occur on the next revolution. Many magnetic tape systems have *read after write*.

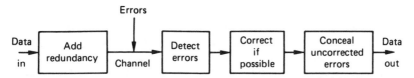

Figure 4.44 The basic stages of an error-correction system. Of these the most critical is the detection stage, since this controls the subsequent actions.

During recording, offtape data are immediately checked for errors. If an error is detected, the tape may abort the recording, reverse to the beginning of the current block and erase it. The data from that block may then be recorded further down the tape. This is the recording equivalent of a retransmission in a communications system.

4.16 Concealment by interpolation

There are some practical differences between data recording for video and the computer data recording application. Although video or audio recorders seldom have time for retries, they have the advantage that there is a certain amount of redundancy in the information conveyed. Thus if an error cannot be corrected, then it can be concealed. If a sample is lost, it is possible to obtain an approximation to it by interpolating between the samples before and after the missing one. Clearly concealment of any kind cannot be used with computer instructions.

If there is too much corruption for concealment, the only course in video is to repeat the previous field or frame in a freeze as it is unlikely that the corrupt picture is watchable.

In general, if use is to be made of concealment on replay, the data must generally be reordered or shuffled prior to recording. To take a simple example, odd-numbered samples are recorded in a different area of the medium from even-numbered samples. On playback, if a gross error occurs on the tape, depending on its position, the result will be either corrupted odd samples or corrupted even samples, but it is most unlikely that both will be lost. Interpolation is then possible if the power of the correction system is exceeded. In practice the shuffle employed in digital video recorders is two-dimensional and rather more complex. Further details can be found in Chapter 6. The concealment technique described here is only suitable for PCM recording. If data reduction has been employed, different concealment techniques will be needed.

It should be stressed that corrected data are indistinguishable from the original and thus there can be no visible or audible artefacts. In contrast, concealment is only an approximation to the original information and could be detectable. In practical video equipment, concealment occurs infrequently unless there is a defect requiring attention, and its presence is difficult to see.

4.17 Modulo-n arithmetic

Conventional arithmetic which is in everyday use relates to the real world of counting actual objects, and to obtain correct answers the concepts of borrow and carry are necessary in the calculations.

There is an alternative type of arithmetic which has no borrow or carry which is known as modulo arithmetic. In modulo-n no number can exceed $n - 1$. If it does, n or whole multiples of n are subtracted until it does not. Thus 25 modulo-16 is 9 and 12 modulo-5 is 2. The count shown in Figure 3.5 is from a 4 bit device which overflows when it reaches 1111 because the carry-out is ignored. If a number of clock pulses m are applied from the zero state, the state of the counter will be given by m mod.16. Thus modulo arithmetic is appropriate to systems in which there is a fixed wordlength and this means that the range of values the

A $\longrightarrow$ $\longrightarrow$ A + B mod. 2 = A $\oplus$ B
B $\longrightarrow$

$$\begin{matrix} 1 \vdots 1 \vdots 0 \ 1 \\ 1 \vdots 1 \vdots 1 \ 0 \end{matrix} \Big\} \xleftarrow{\text{Modulo-2}} \text{sum} \xrightarrow{} \Big\{ \begin{matrix} 1\ 0\ 1\ 0 \\ 1\ 0\ 1\ 0 \end{matrix}$$
$$\overline{0 \vdots 0 \vdots 1\ 1} \qquad\qquad\qquad \overline{0\ 0\ 0\ 0}$$

Each bit position is independently
calculated – no carry

Figure 4.45 In modulo-2 calculations, there can be no carry or borrow operations and conventional addition and subtraction become identical. The XOR gate is a modulo-2 adder.

system can have is restricted by that wordlength. A number range which is restricted in this way is called a finite field.

Modulo-2 is a numbering scheme which is used frequently in digital processes. Figure 4.45 shows that in modulo-2 the conventional addition and subtraction are replaced by the XOR function such that: A + B mod.2 = A XOR B. When multibit values are added mod.2, each column is computed quite independently of any other. This makes mod.2 circuitry very fast in operation as it is not necessary to wait for the carries from lower-order bits to ripple up to the high-order bits.

Modulo-2 arithmetic is not the same as conventional arithmetic and takes some getting used to. For example, adding something to itself in mod.2 always gives the answer zero.

4.18 Parity

The error-detection and error-correction processes are closely related and will be dealt with together here. The actual correction of an error is simplified tremendously by the adoption of binary. As there are only two symbols, 0 and 1, it is enough to know that a symbol is wrong, and the correct value is obvious. Figure 4.46 shows a minimal circuit required for correction once the bit in error has been identified. The XOR (exclusive OR) gate shows up extensively in error correction and the figure also shows the truth table. One way of remembering the characteristics of this useful device is that there will be an output when the inputs

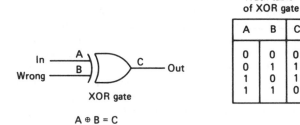

Truth table
of XOR gate

A	B	C
0	0	0
0	1	1
1	0	1
1	1	0

In —— A
Wrong —— B $\rangle$ C —— Out

XOR gate

A $\oplus$ B = C

Figure 4.46 Once the position of the error is identified, the correction process in binary is easy.

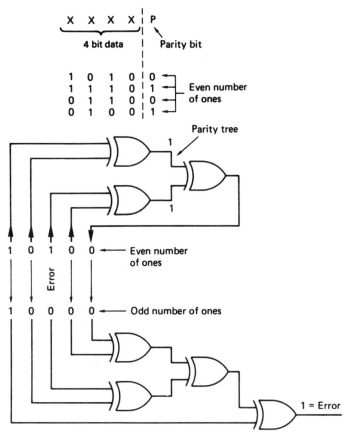

Figure 4.47 Parity checking adds up the number of ones in a word using, in this example, parity trees. One error bit and odd numbers of errors are detected. Even numbers of errors cannot be detected.

are different. Inspection of the truth table will show that there is an even number of ones in each row (zero is an even number) and so the device could also be called an even parity gate. The XOR gate is also a adder in modulo 2.

Parity is a fundamental concept in error detection. In Figure 4.47, the example is given of a 4 bit data word which is to be protected. If an extra bit is added to the word which is calculated in such a way that the total number of ones in the 5 bit word is even, this property can be tested on receipt. The generation of the parity bit in Figure 4.47 can be performed by a number of the ubiquitous XOR gates configured into what is known as a parity tree. In the figure, if a bit is corrupted, the received message will be seen no longer to have an even number of ones. If two bits are corrupted, the failure will be undetected. This example can be used to introduce much of the terminology of error correction. The extra bit added to the message carries no information of its own, since it is calculated from the other bits. It is therefore called a *redundant* bit. The addition of the redundant bit gives the message a special property, i.e. the number of ones is even. A message having some special property *irrespective of the actual data content* is called a *codeword*. All error correction relies on adding redundancy to real data to form codewords for transmission. If any corruption occurs, the intention is that the received message will not have the special property; in other words if the received message is not a codeword there has definitely been an error. The receiver can check for the special property without any prior knowledge of the data content. Thus the same check can be made on all received data. If the received message is a codeword, there probably has not been an error. The word 'probably' must be used because the figure shows that two bits in error will cause the received message to be a codeword, which cannot be discerned from an error-free message. If it is known that generally the only failure mechanism in the channel in question is loss of a single bit, it is *assumed* that receipt of a codeword means that there has been no error. If there is a probability of two error bits, that becomes very nearly the probability of failing to detect an error, since all odd numbers of errors will be detected, and a 4 bit error is much less likely. It is paramount in all error-correction systems that the protection used should be appropriate for the probability of errors to be encountered. An inadequate error-correction system is actually worse than not having any correction. Error correction works by trading probabilities. Error-free performance with a certain error rate is achieved at the expense of performance at higher error rates. Figure 4.48 shows the effect of an error-correction system on the residual BER for a

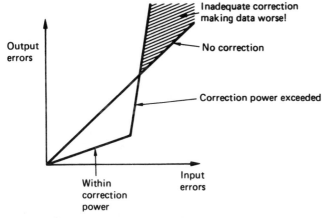

Figure 4.48 An error-correction system can only reduce errors at normal error rates at the expense of increasing errors at higher rates. It is most important to keep a system working to the left of the knee in the graph.

given raw BER. It will be seen that there is a characteristic knee in the graph. If the expected raw BER has been misjudged, the consequences can be disastrous. Another result demonstrated by the example is that we can only guarantee to detect the same number of bits in error as there are redundant bits.

Figure 4.49 shows a strategy known as a crossword code, or product code. The data are formed into a two-dimensional array, in which each location can be a single bit or a multibit symbol. Parity is then generated on both rows and columns. If a single bit or symbol fails, one row parity check and one column parity check will fail, and the failure can be located at the intersection of the two

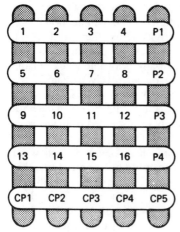

Figure 4.49 A block code. Each location in the block can be a bit or a word. Horizontal parity checks are made by adding P1, P2, etc., and cross-parity or vertical checks are made by adding CP1, CP2, etc. Any symbol in error will be at the intersection of the two failing codewords.

failing checks. Although two symbols in error confuse this simple scheme, using more complex coding in a two-dimensional structure is very powerful, and further examples will be given throughout this chapter.

4.19 Cyclic codes

In digital recording applications, the data are stored serially on a track, and it is desirable to use relatively large data blocks to reduce the amount of the medium devoted to preambles, addressing and synchronizing. The principle of codewords having a special characteristic will still be employed, but they will be generated and checked algorithmically by equations. The syndrome will then be converted to the bit(s) in error by solving equations.

Where data can be accessed serially, simple circuitry can be used because the same gate will be used for many XOR operations. The circuit of Figure 4.50 is a kind of shift register, but with a particular feedback arrangement which leads it to be known as a twisted-ring counter. If seven message bits A–G are applied serially to this circuit, and each one of them is clocked, the outcome can be followed in the diagram. As bit A is presented and the system is clocked, bit A will enter the left-hand latch. When bits B and C are presented, A moves across to the right. Both XOR gates will have A on the upper input from the right-hand latch, the left one has D on the lower input and the right one has B on the lower input. When clocked, the left latch will thus be loaded with the XOR of A and D, and the right one with the XOR of A and B. The remainder of the sequence can be followed, bearing in mind that when the same term appears on both inputs of an XOR gate, it goes out, as the exclusive OR of something with itself is nothing. At the end of the process, the latches contain three different expressions. Essentially, the circuit makes three parity checks through the message, leaving the result of each in the three stages of the register. In the figure, these expressions have been used to draw up a check matrix. The significance of these

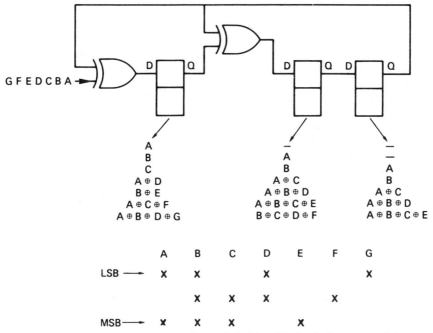

Figure 4.50 When seven successive bits A–G are clocked into this circuit, the contents of the three latches are shown for each clock. The final result is a parity-check matrix.

steps can now be explained. The bits A B C and D are four data bits, and the bits E F and G are redundancy. When the redundancy is calculated, bit E is chosen so that there are an even number of ones in bits A B C and E; bit F is chosen such that the same applies to bits B C D and F, and similarly for bit G. Thus the four data bits and the three check bits form a 7 bit codeword. If there is no error in the codeword, when it is fed into the circuit shown, the result of each of the three parity checks will be zero and every stage of the shift register will be cleared. As the register has eight possible states, and one of them is the error-free condition, then there are seven remaining states; hence the seven bit codeword. If a bit in the codeword is corrupted, there will be a non-zero result. For example, if bit D fails, the check on bits A B D and G will fail, and a one will appear in the left-hand latch. The check on bits B C D F will also fail, and the centre latch will set. The check on bits A B C E will not fail, because D is not involved in it, making the right-hand bit zero. There will be a syndrome of 011 in the register, and this will be seen from the check matrix to correspond to an error in bit D. Whichever bit fails, there will be a different 3 bit syndrome which uniquely identifies the failed bit. As there are only three latches, there can be eight different syndromes. One of these is zero, which is the error-free condition, and so there are seven remaining error syndromes. The length of the codeword cannot exceed seven bits, or there would not be enough syndromes to correct all of the bits. This can also be made to tie in with the generation of the check matrix. If fourteen bits, A to N, were fed into the circuit shown, the result would be that the check matrix repeated twice, and if a syndrome of 011 were to result, it could not be determined whether bit D or bit K failed. Because the check repeats every 7 bits, the code is said to be a cyclic redundancy check (CRC) code.

It has been seen that the circuit shown makes a matrix check on a received word to determine if there has been an error, but the same circuit can also be used to generate the check bits. To visualize how this is done, examine what happens if only the data bits A B C and D are known, and the check bits E F and G are set to zero. If this message, ABCD000, is fed into the circuit, the left-hand latch will afterwards contain the XOR of A B C and zero, which is of course what E should be. The centre latch will contain the XOR of B C D and zero, which is what F should be and so on. This process is not quite ideal, however, because it is necessary to wait for three clock periods after entering the data before the check bits are available. Where the data are simultaneously being recorded and fed into the encoder, the delay would prevent the check bits being easily added to the end of the data stream. This problem can be overcome by slightly modifying the encoder circuit as shown in Figure 4.51. By moving the position

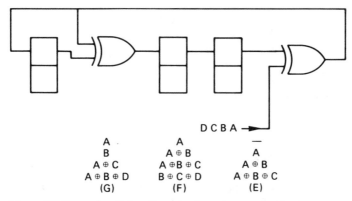

A	A	—
B	A ⊕ B	A
A ⊕ C	A ⊕B ⊕ C	A ⊕ B
A ⊕ B ⊕ D	B ⊕ C ⊕ D	A ⊕ B ⊕ C
(G)	(F)	(E)

DCBA →

Figure 4.51 By moving the insertion point three places to the right, the calculation of the check bits is completed in only four clock periods and they can follow the data immediately. This is equivalent to premultiplying the data by x^3.

of the input to the right, the operation of the circuit is advanced so that the check bits are ready after only four clocks. The process can be followed in the diagram for the four data bits A B C and D. On the first clock, bit A enters the left two latches, whereas on the second clock, bit B will appear on the upper input of the left XOR gate, with bit A on the lower input, causing the centre latch to load the XOR of A and B and so on.

The way in which the cyclic codes work has been described in engineering terms, but it can be described mathematically if analysis is contemplated.

Just as the position of a decimal digit in a number determines the power of ten (whether that digit means one, ten or a hundred), the position of a binary digit determines the power of two (whether it means one, two or four). It is possible to rewrite a binary number so that it is expressed as a list of powers of two. For example, the binary number 1101 means 8 + 4 + 1, and can be written:

$$2^3 + 2^2 + 2^0$$

In fact, much of the theory of error correction applies to symbols in number bases other than 2, so that the number can also be written more generally as

$$x^3 + x^2 + 1 \ (2^0 = 1)$$

which also looks much more impressive. This expression, containing as it does various powers, is of course a polynomial, and the circuit of Figure 4.50 which has been seen to construct a parity-check matrix on a codeword can also be described as calculating the remainder due to dividing the input by a polynomial using modulo-2 arithmetic. In modulo-2 there are no borrows or carries, and addition and subtraction are replaced by the XOR function, which makes hardware implementation very easy. In Figure 4.52 it will be seen that the circuit of Figure 4.50 actually divides the codeword by a polynomial which is:

$$x^3 + x + 1 \text{ or } 1011$$

This can be deduced from the fact that the right-hand bit is fed into two lower-order stages of the register at once. Once all the bits of the message have been clocked in, the circuit contains the remainder. In mathematical terms, the special property of a codeword is that it is a polynomial which yields a remainder of zero when divided by the generating polynomial. The receiver will make this division, and the result should be zero in the error-free case. Thus the codeword itself disappears from the division. If an error has occurred it is considered that this is due to an error polynomial which has been added to the codeword polynomial. If a codeword divided by the check polynomial is zero, a non-zero syndrome must represent the error polynomial divided by the check polynomial. Thus if the syndrome is multiplied by the check polynomial, the latter will be cancelled out and the result will be the error polynomial. If this is added modulo-2 to the received word, it will cancel out the error and leave the corrected data.

Some examples of modulo-2 division are given in Figure 4.52 which can be compared with the parallel computation of parity checks according to the matrix of Figure 4.50.

The process of generating the codeword from the original data can also be described mathematically. If a codeword has to give zero remainder when divided, it follows that the data can be converted to a codeword by adding the remainder when the data are divided. Generally speaking the remainder would have to be subtracted, but in modulo-2 there is no distinction. This process is also illustrated in Figure 4.52. The four data bits have three zeros placed on the right-hand end, to make the wordlength equal to that of a codeword, and this word is then divided by the polynomial to calculate the remainder. The remainder is added to the zero-extended data to form a codeword. The modified circuit of Figure 4.51 can be described as premultiplying the data by x^3 before dividing.

CRC codes are of primary importance for detecting errors, and several have been standardized for use in digital communications. The most common of these are:

$$x^{16} + x^{15} + x^2 + 1 \text{ (CRC-16)}$$

$$x^{16} + x^{12} + x^5 + 1 \text{ (CRC-CCITT)}$$

Cyclic codes are used for signature analysis, which is a reliability testing technique described in Chapter 5. A signature is a parameter similar to a CRCC, but which is used in a slightly different way. Figure 4.53 shows that the concept

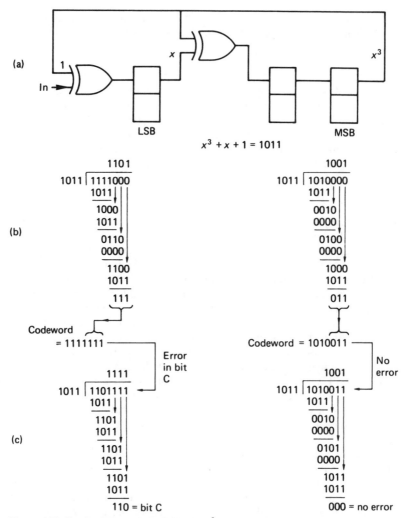

Figure 4.52 Circuit of Figure 4.50 divides by $x^3 + x + 1$ to find remainder. In (b) this is used to calculate check bits. In (c) right, zero syndrome, no error.

of signature analysis is based on dividing data blocks by a polynomial. If a given fixed digital test pattern is produced by the generator of Figure 4.53, it will result in a fixed remainder when divided by a polynomial on the left. The remainder is the signature. If the same signature is reliably obtained when the polynomial divider is moved to the far end of a transmission system, then the transmission system is error free. It is a characteristic of cyclic codes that a single bit in error in a lengthy message will change the signature so that it can be detected. In practice, the data blocks in digital video are quite large, and will exceed the length of the codeword. The cyclic code may repeat several times within the block. Thus location of errors is impossible, but this is not a goal of signature analysis as it is not an error-correction process.

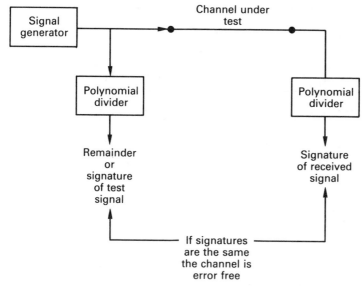

Figure 4.53 In signature analysis a data block is divided by a polynomial to obtain a remainder or signature. If the same signature is obtained elsewhere, the occurrence of an error is remote.

4.20 The Galois field

Figure 4.54 shows a simple circuit consisting of three D-type latches which are clocked simultaneously. They are connected in series to form a shift register. At (a) a feedback connection has been taken from the output to the input and the result is a ring counter where the bits contained will recirculate endlessly. At (b) one XOR gate is added so that the output is fed back to more than one stage. The result is known as a twisted-ring counter and it has some interesting properties. Whenever the circuit is clocked, the left-hand bit moves to the right-hand latch, the centre bit moves to the left-hand latch and the centre latch becomes the XOR of the two outer latches. The figure shows that whatever the starting condition of the three bits in the latches, the same state will always be reached again after seven clocks, except if zero is used. The states of the latches form an endless ring of non-sequential numbers called a Galois field after the French mathematical prodigy Evariste Galois who discovered them. The states of the circuit form a maximum length sequence because there are as many states as are permitted by the wordlength. As the states of the sequence have many of the characteristics of random numbers, yet are repeatable, the result can also be called a pseudo-random sequence (prs). As the all-zeros case is disallowed, the length of a maximum length sequence generated by a register of m bits cannot exceed (2^m-1) states. The Galois field, however includes the zero term. It is useful to explore the bizarre mathematics of Galois fields which use modulo-2 arithmetic. Familiarity with such manipulations is helpful when studying the error correction, particularly the Reed–Solomon codes. They will also be found in processes which require pseudo-random numbers such as randomized channel codes.

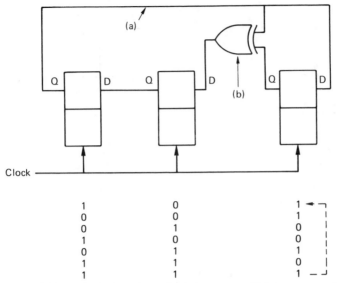

Figure 4.54 The circuit shown is a twisted-ring counter which has an unusual feedback arrangement. Clocking the counter causes it to pass through a series of non-sequential values. See text for details.

The circuit of Figure 4.54 can be considered as a counter and the four points shown will then be representing different powers of 2 from the MSB on the left to the LSB on the right. The feedback connection from the MSB to the other stages means that whenever the MSB becomes one, two other powers are also forced to one so that the code of 1011 is generated.

Each state of the circuit can be described by combinations of powers of x, such as:

$$x^2 = 100$$

$$x = 010$$

$$x^2 + x = 110, \text{ etc.}$$

The fact that three bits have the same state because they are connected together is represented by the mod.2 equation:

$$x^3 + x + 1 = 0$$

Let $x = a$, which is a primitive element. Now:

$$a^3 + a + 1 = 0 \tag{4.1}$$

In modulo-2:

$$a + a = a^2 + a^2 = 0$$

$$a = x = 010$$

$$a^2 = x^2 = 100$$

$$a^3 = a + 1 = 011 \text{ from (4.1)}$$

$$a^4 = a \times a^3 = a(a + 1) = a^2 + a = 110$$

$$a^5 = a^2 + a + 1 = 111$$

$$a^6 = a \times a^5 = a(a^2 + a + 1)$$

$$= a^3 + a^2 + a = a + 1 + a^2 + a$$

$$= a^2 + 1 = 101$$

$$a^7 = a(a^2 + 1) = a^3 + a$$

$$a + 1 + a = 1 = 001$$

In this way it can be seen that the complete set of elements of the Galois field can be expressed by successive powers of the primitive element. Note that the twisted-ring circuit of Figure 4.54 simply raises a to higher and higher powers as it is clocked; thus the seemingly complex multibit changes caused by a single clock of the register become simple to calculate using the correct primitive and the appropriate power.

The numbers produced by the twisted-ring counter are not random; they are completely predictable if the equation is known. However, the sequences produced are sufficiently similar to random numbers that in many cases they will be useful. They are thus referred to as pseudo-random sequences. The feedback connection is chosen such that the expression it implements will not factorize. Otherwise a maximum-length sequence could not be generated because the circuit might sequence around one or other of the factors depending on the initial condition. A useful analogy is to compare the operation of a pair of meshed gears. If the gears have a number of teeth which is relatively prime, many revolutions are necessary to make the same pair of teeth touch again. If the number of teeth have a common multiple, far fewer turns are needed.

4.21 Introduction to the Reed–Solomon codes

The Reed–Solomon codes (Irving Reed and Gustave Solomon) are inherently burst correcting[15] because they work on multibit symbols rather than individual bits. The R–S codes are also extremely flexible in use. One code may be used both to detect and correct errors and the number of bursts which are correctable can be chosen at the design stage by the amount of redundancy. A further advantage of the R–S codes is that they can be used in conjunction with a separate error-detection mechanism in which case they perform only the correction by erasure. R–S codes operate at the theoretical limit of correcting efficiency. In other words, no more efficient code can be found.

In the simple CRC system described in Section 4.18, the effect of the error is detected by ensuring that the codeword can be divided by a polynomial. The CRC codeword was created by adding a redundant symbol to the data. In the Reed–Solomon codes, several errors can be isolated by ensuring that the codeword will divide by a number of polynomials. Clearly if the codeword must divide by, say, two polynomials, it must have two redundant symbols. This is the minimum case of an R–S code. On receiving an R–S coded message there will be two syndromes following the division. In the error-free case, these will both be zero. If both are not zero, there is an error.

It has been stated that the effect of an error is to add an error polynomial to the message polynomial. The number of terms in the error polynomial is the same as the number of errors in the codeword. The codeword divides to zero and the syndromes are a function of the error only. There are two syndromes and two equations. By solving these simultaneous equations it is possible to obtain two unknowns. One of these is the position of the error, known as the *locator*, and the other is the error bit pattern, known as the *corrector*. As the locator is the same size as the code symbol, the length of the codeword is determined by the size of the symbol. A symbol size of 8 bits is commonly used because it fits in conveniently with both 16 bit audio samples and byte-oriented computers. An 8 bit syndrome results in a locator of the same wordlength; 8 bits have 2^8 combinations, but one of these is the error-free condition, and so the locator can specify one of only 255 symbols. As each symbol contains 8 bits, the codeword will be $255 \times 8 = 2040$ bits long.

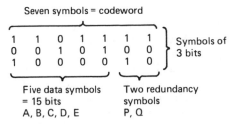

Figure 4.55 A Reed–Solomon codeword. As the symbols are of 3 bits, there can only be eight possible syndrome values. One of these is all zeros, the error-free case, and so it is only possible to point to seven errors; hence the codeword length of seven symbols. Two of these are redundant, leaving five data symbols.

As further examples, 5 bit symbols could be used to form a codeword 31 symbols long, and 3 bit symbols would form a codeword seven symbols long. This latter size is small enough to permit some worked examples, and will be used further here. Figure 4.55 shows that in the seven-symbol codeword, five symbols of three bits each, A–E, are the data, and P and Q are the two redundant symbols. This simple example will locate and correct a single symbol in error. It does not matter, however, how many bits in the symbol are in error.

The two check symbols are solutions to the following equations:

$$A \oplus B \oplus C \oplus D \oplus E \oplus P \oplus Q = 0$$

$$a^7A \oplus a^6B \oplus a^5C \oplus a^4D \oplus a^3E \oplus a^2P \oplus aQ = 0$$

where a is a constant. The original data A–E followed by the redundancy P and Q pass through the channel.

The receiver makes two checks on the message to see if it is a codeword. This is done by calculating syndromes using the following expressions, where the prime implies the received symbol which is not necessarily correct:

$$S_0 = A' \oplus B' \oplus C' \oplus D' \oplus E' \oplus P' \oplus Q'$$

(this is in fact a simple parity check)

$$S_1 = a^7A' \oplus a^6B' \oplus a^5C' \oplus a^4D' \oplus a^3E' \oplus a^2P' \oplus aQ'$$

If two syndromes of all zeros are not obtained, there has been an error. The information carried in the syndromes will be used to correct the error. For the purpose of illustration, let it be considered that D' has been corrupted before moving to the general case. D' can be considered to be the result of adding an error of value E to the original value D such that $D' = D \oplus E$.

$$A \oplus B \oplus C \oplus D \oplus E \oplus P \oplus Q = 0$$

then:

$$A \oplus B \oplus C \oplus (D \oplus E) \oplus E \oplus P \oplus Q = E = S_0$$

As:

$$D' = D \oplus E$$

then:

$$D = D' \oplus E = D' \oplus S_0$$

Thus the value of the corrector is known immediately because it is the same as the parity syndrome S_0. The corrected data symbol is obtained simply by adding S_0 to the incorrect symbol.

At this stage, however, the corrupted symbol has not yet been identified, but this is equally straightforward. As:

$$a^7A \oplus a^6B \oplus a^5C \oplus a^4D \oplus a^3E \oplus a^2P \oplus aQ = 0$$

then:

$$a^7A \oplus a^6B \oplus a^5C \oplus a^4(D \oplus E) \oplus a^3E \oplus a^2P \oplus aQ = a^4E = S_1$$

Thus the syndrome S_1 is the error bit pattern E, but it has been raised to a power of a which is a function of the position of the error symbol in the block. If the position of the error is in symbol k, then k is the locator value and:

$$S_0 \times a^k = S_1$$

Hence:

$$a^k = \frac{S_1}{S_0}$$

The value of k can be found by multiplying S_0 by various powers of a until the product is the same as S_1. Then the power of a necessary is equal to k. The use of the descending powers of a in the codeword calculation is now clear because the error is then multiplied by a different power of a dependent upon its position. S_1 is known as the locator, because it gives the position of the error. The process of finding the error position by experiment is known as a Chien search.

Whilst the expressions above show that the values of P and Q are such that the two syndrome expressions sum to zero, it is not yet clear how P and Q are calculated from the data. Expressions for P and Q can be found by solving the two R–S equations simultaneously. This has been done in Appendix 4.1. The following expressions must be used to calculate P and Q from the data in order to satisfy the codeword equations. These are:

$$P = a^6A \oplus aB \oplus a^2C \oplus a^5D \oplus a^3E$$
$$Q = a^2A \oplus a^3B \oplus a^6C \oplus a^4D \oplus aE$$

In both the calculation of the redundancy shown here and the calculation of the corrector and the locator it is necessary to perform numerous multiplications and raising to powers. This appears to present a formidable calculation problem at both the encoder and the decoder. This would be the case if the calculations involved were conventionally executed. However, the calculations can be simplified by using logarithms. Instead of multiplying two numbers, their logarithms are added. In order to find the cube of a number, its logarithm is added three times. Division is performed by subtracting the logarithms. Thus all of the manipulations necessary can be achieved with addition or subtraction, which is straightforward in logic circuits.

The success of this approach depends upon simple implementation of log tables. As was seen in Section 4.20, raising a constant, a, known as the *primitive element* to successively higher powers in modulo-2 gives rise to a Galois field. Each element of the field represents a different power n of a. It is a fundamental of the R–S codes that all of the symbols used for data, redundancy and syndromes are considered to be elements of a Galois field. The number of bits in the symbol determines the size of the Galois field, and hence the number of symbols in the codeword.

In Figure 4.56, the binary values of the elements are shown alongside the power of a they represent. In the R–S codes, symbols are no longer considered simply as binary numbers, but also as equivalent powers of a. In Reed–Solomon coding and decoding, each symbol will be multiplied by some power of a. Thus if the symbol is also known as a power of a it is only necessary to add the two powers. For example, if it is necessary to multiply the data symbol 100 by a^3, the calculation proceeds as follows, referring to Figure 4.56.

$$100 = a^2 \text{ so } 100 \times a^3 = a^{(2+3)} = a^5 = 111$$

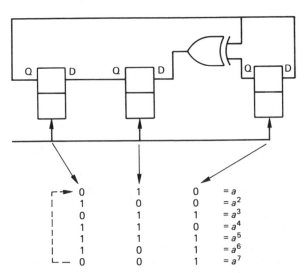

Figure 4.56 The bit patterns of a Galois field expressed as powers of the primitive element a. This diagram can be used as a form of log table in order to multiply binary numbers. Instead of an actual multiplication, the appropriate powers of a are simply added.

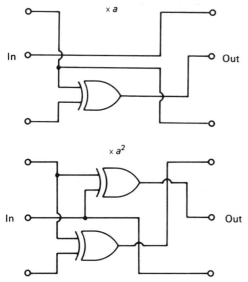

Figure 4.57 Some examples of GF multiplier circuits.

Note that the results of a Galois multiplication are quite different from binary multiplication. Because all products must be elements of the field, sums of powers which exceed seven wrap around by having seven subtracted. For example:

$$a^5 \times a^6 = a^{11} = a^4 = 110$$

Figure 4.57 shows some examples of circuits which will perform this kind of multiplication. Note that they require a minimum amount of logic.

Input data	A	101	$a^6 A = 111$	$a^2 A = 010$	
	B	100	$a\,B = 011$	$a^3 B = 111$	
	C	010	$a^2 C = 011$	$a^6 C = 001$	
	D	100	$a^5 D = 001$	$a^4 D = 101$	
	E	111	$a^3 E = 010$	$a\,E = 101$	
Check symbols	P	100 ← ———— 100		100	
	Q	100 ← ————————————			

Codeword	A	101	$a^7 A = 101$
	B	100	$a^6 B = 010$
	C	010	$a^5 C = 101$
	D	100	$a^4 D = 101$
	E	111	$a^3 E = 010$
	P	100	$a^2 P = 110$
	Q	100	$a\,Q = 011$
	$S_0 = \overline{000}$		$S_1 = \overline{000}$ ← —— Both syndromes zero

Figure 4.58 Five data symbols A–E are used as terms in the generator polynomials derived in Appendix 4.1 to calculate two redundant symbols P and Q. An example is shown at the top. Below is the result of using the codeword symbols A–Q as terms in the checking polynomials. As there is no error, both syndromes are zero.

A, B, C, D, E

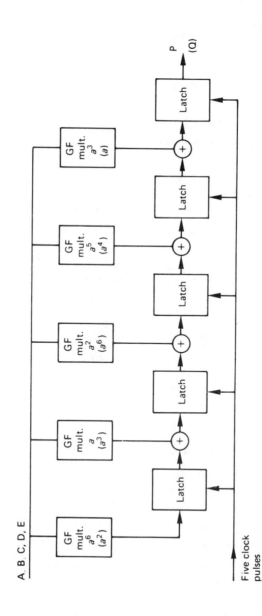

Five clock
pulses

Figure 4.59 If the five data symbols of Figure 4.58 are supplied to this circuit in sequence, after five clocks one of the check symbols will appear at the output. Terms without brackets will calculate P, bracketed terms calculate Q.

Figure 4.58 shows an example of the Reed–Solomon encoding process. The Galois field shown in Figure 4.56 has been used, having the primitive element a = 010. At the beginning of the calculation of P, the symbol A is multiplied by a^6. This is done by converting A to a power of a. According to Figure 4.56, 101 = a^6 and so the product will be $a^{(6+6)} = a^{12} = a^5 = 111$. In the same way, B is multiplied by a, and so on, and the products are added modulo-2. A similar process is used to calculate Q.

Figure 4.59 shows a circuit which can calculate P or Q. The symbols A–E are presented in succession, and the circuit is clocked for each one. On the first clock, a^6A is stored in the left-hand latch. If B is now provided at the input, the second GF multiplier produces aB and this is added to the output of the first latch and when clocked will be stored in the second latch which now contains a^6A + aB. The process continues in this fashion until the complete expression for P is available in the right-hand latch. The intermediate contents of the right-hand latch are ignored.

The entire codeword now exists, and can be recorded or transmitted. Figure 4.58 also demonstrates that the codeword satisfies the checking equations. The modulo-2 sum of the seven symbols, S_0, is 000 because each column has an even number of ones. The calculation of S_1 requires multiplication by descending powers of a. The modulo-2 sum of the products is again zero. These calculations confirm that the redundancy calculation was properly carried out.

Figure 4.60 gives three examples of error correction based on this codeword. The erroneous symbol is marked with a dash. As there has been an error, the syndromes S_0 and S_1 will not be zero.

7	A	101	a^7 A = 101	
6	B	100	a^6 B = 010	$\dfrac{S_1}{S_0} = \dfrac{a^4}{1} = a^4$
5	C	010	a^5 C = 101	
4	D'	101	a^4 D' = 011 — $k = 4$
3	E	111	a^3 E = 010	
2	P	100	a^2 P = 110	D' + S_0 = 101 + 001
1	Q	100	a Q = 011	D = 100
	S_0 =	001	S_1 = 110	

7	A	101	a^7 A = 101	
6	B	100	a^6 B = 010	$\dfrac{S_1}{S_0} = \dfrac{1}{a^2} = \dfrac{1}{a^2} \times \dfrac{a^5}{a^5} = a^5$
5	C'	110	a^5 C = 100 —	
4	D	100	a^4 D = 101 — $k = 5$
3	E	111	a^3 E = 010	
2	P	100	a^2 P = 110	C' + S_0 = 110 + 100
1	Q	100	a Q = 011	C = 010
	S_0 =	100	S_1 = 001	

7	A'	111	a^7 A = 111	
6	B	100	a^6 B = 010	$\dfrac{S_1}{S_0} = \dfrac{a}{a} = 001 = a^7$
5	C	010	a^5 C = 101	
4	D	100	a^4 D = 101	$k = 7$
3	E	111	a^3 E = 010	
2	P	100	a^2 P = 110	A' + S_0 = 111 + 010
1	Q	100	a Q = 011	A = 101
	S_0 =	010	S_1 = 010	

Figure 4.60 Three examples of error location and correction. The number of bits in error in a symbol is irrelevant; if all three were wrong, S_0 would be 111, but correction is still possible.

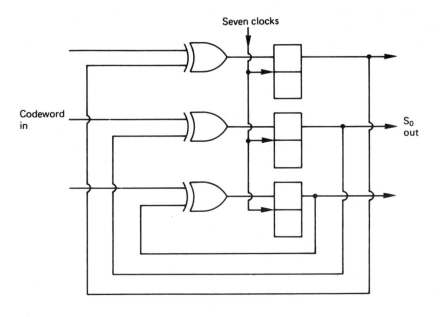

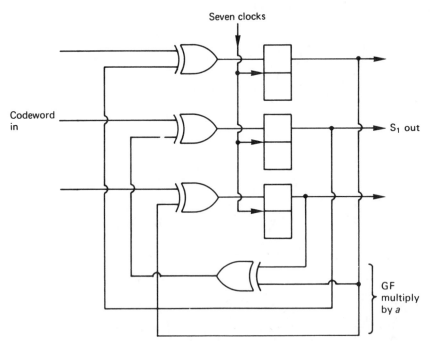

Figure 4.61 Circuits for parallel calculation of syndromes S_0, S_1. S_0 is a simple parity check. S_1 has a GF multiplication by a in the feedback, so that A is multiplied by a^7, B is multiplied by a^6, etc., and all are summed to give S_1.

Figure 4.61 shows circuits suitable for parallel calculation of the two syndromes at the receiver. The S_0 circuit is a simple parity checker which accumulates the modulo-2 sum of all symbols fed to it. The S_1 circuit is more subtle, because it contains a Galois field (GF) multiplier in a feedback loop, such that early symbols fed in are raised to higher powers than later symbols because they have been recirculated through the GF multiplier more often. It is possible to compare the operation of these circuits with the example of Figure 4.60 and with subsequent examples to confirm that the same results are obtained.

A	101	$a^7 A =$	101
B	100	$a^6 B =$	010
$(C \oplus E_C)$	001	$a^5 (C \oplus E_C)$	111
$(D \oplus E_D)$	010	$a^4 (D \oplus E_D)$	111
E	111	$a^3 E =$	010
P	100	$a^2 P =$	110
Q	100	$a Q =$	011
$S_1 =$	101	$S_1 =$	000

$S_0 = E_C \oplus E_D \qquad S_1 = a^5 E_C \oplus a^4 E_D$

$S_1 = a^5 E_C \oplus a^4 (S_0 \oplus E_C)$

$\quad = a^5 E_C \oplus a^4 S_0 \oplus a^4 E_C$

$\therefore E_C = \dfrac{S_1 \oplus a^4 S_0}{a^5 \oplus a^4} = \dfrac{000 \oplus 011}{001} = 011$

$C = (C \oplus E_C) \oplus E_C = 001 \oplus 011 = 010$

$S_1 = a^5 (S_0 \ominus E_D) \oplus a^4 E_D$

$\quad = a^5 S_0 \oplus a^5 E_D \oplus a^4 E_D$

$\therefore E_D = \dfrac{S_1 \oplus a^5 S_0}{a^5 \oplus a^4} = \dfrac{000 \oplus 110}{001} = 110$

$D = (D \oplus E_D) + E_D = 010 \oplus 110 = 100$ **(a)**

A	101	$a^7 A = 101$	
B	100	$a^6 B = 010$	$S_0 = C \oplus D$
C	000	$a^5 C = 000$	
D	000	$a^4 D = 000$	$S_1 = a^5 C \oplus a^4 D$
E	111	$a^3 E = 010$	
P	100	$a^2 P = 110$	
Q	100	$a Q = 011$	
$S_0 = 100$		$S_1 = 000$	

$S_1 = a^5 S_0 \oplus a^5 D \oplus a^4 D = a^5 S_0 \oplus D$

$\therefore D = S_1 \oplus a^5 S_0 = 000 \oplus 100 = 100$

$S_1 = a^5 C \oplus a^4 C \oplus a^4 S_0 = C \oplus a^4 S_0$

$\therefore C = S_1 \oplus a^4 S_0 = 000 \oplus 010 = 010$

(b)

Figure 4.62 If the location of errors is known, then the syndromes are a known function of the two errors as shown in (a). It is, however, much simpler to set the incorrect symbols to zero, i.e. to *erase* them as in (b). Then the syndromes are a function of the wanted symbols and correction is easier.

4.22 Correction by erasure

In the examples of Figure 4.60, two redundant symbols P and Q have been used to locate and correct one error symbol. If the positions of errors are known by some separate mechanism (see product codes, Section 4.24) the locator need not be calculated. The simultaneous equations may instead be solved for two correctors. In this case the number of symbols which can be corrected is equal to the number of redundant symbols. In Figure 4.62(a) two errors have taken place, and it is known that they are in symbols C and D. Since S_0 is a simple parity check, it will reflect the modulo-2 sum of the two errors. Hence:

$$S_0 = E_C \oplus E_D$$

The two errors will have been multiplied by different powers in S_1, such that:

$$S_1 = a^5 E_C \oplus a^4 E_D$$

These two equations can be solved, as shown in the figure, to find E_C and E_D, and the correct value of the symbols will be obtained by adding these correctors to the erroneous values. It is, however, easier to set the values of the symbols in error to zero. In this way the nature of the error is rendered irrelevant and it does not enter the calculation. This setting of symbols to zero gives rise to the term erasure. In this case:

$$S_0 = C \oplus D$$

$$S_1 = aC^5 + aD^4$$

Erasing the symbols in error makes the errors equal to the correct symbol values and these are found more simply as shown in Figure 4.62(b).

Practical systems will be designed to correct more symbols in error than in the simple examples given here. If it is proposed to correct by erasure an arbitrary number of symbols in error given by t, the codeword must be divisible by t different polynomials. Alternatively if the errors must be located and corrected, $2t$ polynomials will be needed. These will be of the form $(x + a^n)$ where n takes all values up to t or $2t$. a is the primitive element discussed in Section 4.2.

Where four symbols are to be corrected by erasure, or two symbols are to be located and corrected, four redundant symbols are necessary, and the codeword polynomial must then be divisible by:

$$(x + a^0)(x + a^1)(x + a^2)(x + a^3)$$

Upon receipt of the message, four syndromes must be calculated, and the four correctors or the two error patterns and their positions are determined by solving four simultaneous equations. This generally requires an iterative procedure, and a number of algorithms have been developed for the purpose.[16-18] Modern DVTR formats use 8 bit R–S codes and erasure extensively. The primitive polynomial commonly used with GF(256) is:

$$x^8 + x^4 + x^3 + x^2 + 1$$

The codeword will be 255 bytes long but will often be shortened by puncturing. The larger Galois fields require less redundancy, but the computational problem increases. LSI chips have been developed specifically for R–S decoding in many high-volume formats.

4.23 Interleaving

The concept of bit interleaving was introduced in connection with a single-bit correcting code to allow it to correct small bursts. With burst-correcting codes such as Reed–Solomon, bit interleave is unnecessary. In most channels, particularly high-density recording channels used for digital video or audio, the burst size may be many bytes rather than bits, and to rely on a code alone to correct such errors would require a lot of redundancy. The solution in this case is to employ symbol interleaving, as shown in Figure 4.63. Several codewords are encoded from input data, but these are not recorded in the order they were input, but are physically reordered in the channel, so that a real burst error is split

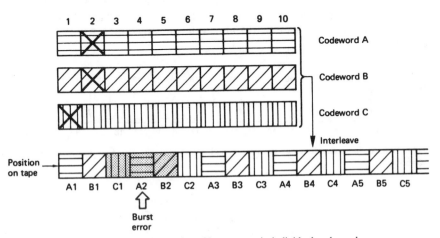

Figure 4.63 The interleave controls the size of burst errors in individual codewords.

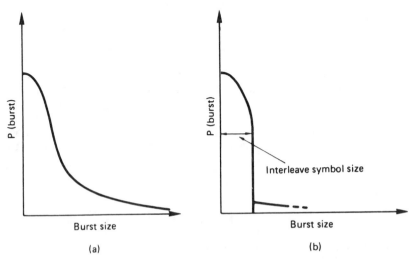

Figure 4.64 (a) The distribution of burst sizes might look like this. (b) Following interleave, the burst size within a codeword is controlled to that of the interleave symbol size, except for gross errors which have low probability.

into smaller bursts in several codewords. The size of the burst seen by each codeword is now determined primarily by the parameters of the interleave, and Figure 4.64 shows that the probability of occurrence of bursts with respect to the burst length in a given codeword is modified. The number of bits in the interleave word can be made equal to the burst-correcting ability of the code in the knowledge that it will be exceeded only very infrequently.

There are a number of different ways in which interleaving can be performed. Figure 4.65 shows that in block interleaving, words are reordered within blocks which are themselves in the correct order. This approach is attractive for rotary-head recorders, because the scanning process naturally divides the tape up into blocks. The block interleave is achieved by writing samples into a memory in sequential address locations from a counter, and reading the memory with non-sequential addresses from a sequencer. The effect is to convert a one-dimensional sequence of samples into a two-dimensional structure having rows and columns.

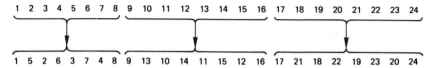

Figure 4.65 In block interleaving, data are scrambled within blocks which are themselves in the correct order.

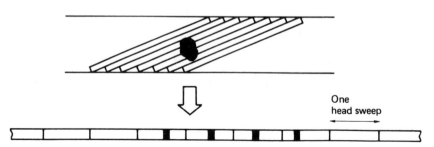

Figure 4.66 Helical-scan recorders produce a form of mechanical interleaving, because one large defect on the medium becomes distributed over several head sweeps.

Rotary-head recorders naturally interleave spatially on the tape. Figure 4.66 shows that a single large tape defect becomes a series of small defects owing to the geometry of helical scanning.

4.24 Product codes

In the presence of burst errors alone, the system of interleaving works very well, but it is known that in most practical channels there are also uncorrelated errors of a few bits due to noise. Figure 4.67 shows an interleaving system where a dropout-induced burst error has occurred which is at the maximum correctable size. All three codewords involved are working at their limit of one symbol. A random error due to noise in the vicinity of a burst error will cause the correction

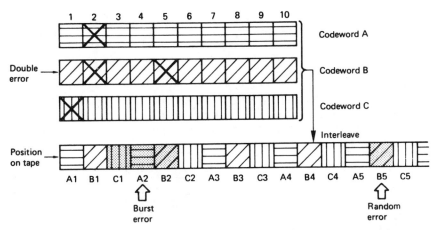

Figure 4.67 The interleave system falls down when a random error occurs adjacent to a burst.

power of the code to be exceeded. Thus a random error of a single bit causes a further entire symbol to fail. This is a weakness of an interleave solely designed to handle dropout-induced bursts. Practical high-density equipment must address the problem of noise-induced or random errors and burst errors ocurring at the same time. This is done by forming codewords both before and after the interleave process. In block interleaving, this results in a *product code*, whereas in the case of convolutional interleave the result is called *cross-interleaving*.

Figure 4.68 shows that in a product code the redundancy calculated first and checked last is called the outer code, and the redundancy calculated second and checked first is called the inner code. The inner code is formed along tracks on the medium. Random errors due to noise are corrected by the inner code and do not impair the burst-correcting power of the outer code. Burst errors are declared uncorrectable by the inner code which flags the bad samples on the way into the de-interleave memory. The outer code reads the error flags in order to correct the flagged symbols by erasure. The error flags are also known as erasure flags. As it does not have to compute the error locations, the outer code needs half as much redundancy for the same correction power. Thus the inner code redundancy does not raise the code overhead. The combination of codewords with interleaving in several dimensions yields an error-protection strategy which is truly synergistic, in that the end result is more powerful than the sum of the parts. Needless to say, the technique is used extensively in modern DVTR formats.

4.25 Introduction to error correction in D-3

The interleave and error-correction systems of D-3 will now be discussed by way of a representative example. Figure 4.69 shows a conceptual block diagram of the system which shows that D-3 uses a product code formed by producing Reed–Solomon codewords at right angles across an array. The array is formed in a memory which contains an entire field of data, and the layout used in PAL can be seen in Figure 4.70. Incoming samples (bytes) are written into the array in columns. Each column is then made into a code word by the addition of 8 bytes of redundancy. These are the outer code words. Each row of the array is then

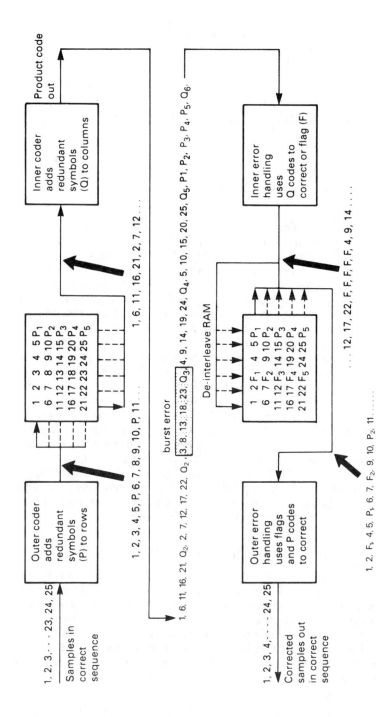

Figure 4.68 In addition to the redundancy P on rows, inner redundancy Q is also generated on columns. On replay, the Q code checker will pass on flags F if it finds an error too large to handle itself. The flags pass through the de-interleave process and are used by the outer error correction to identify which symbol in the row needs correcting with P redundancy. The concept of crossing two codes in this way is called a product code.

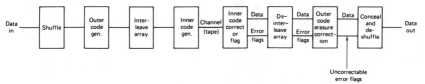

Figure 4.69 Block diagram of error-correction strategy of D-3 DVTR.

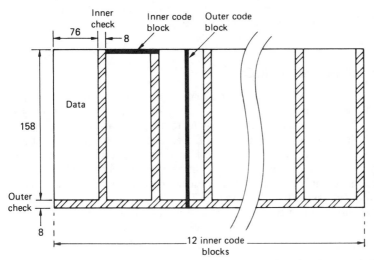

Figure 4.70 The field array of PAL D-3. Incoming data are written in columns and eight R–S bytes are added to make outer codes. The array is split into 12 equal parts, and one row of each part has eight R–S bytes added to make an inner code. There are thus 12 inner code blocks across the array. One inner code is held in one sync block and in practice the codeword is extended to protect the ID pattern.

formed into eight inner code words by the addition of 8 bytes of redundancy to each. In order to make a recording, the memory is read in rows, and one inner codeword fits into one sync block along the track. Every sync block in an eight-field sequence is given a unique header address. This process continues until the entire contents of the array has been laid on tape. Owing to the use of segmentation this will require three segments in NTSC and four in PAL.

Writing the memory in columns and reading in rows achieves the necessary interleave.

On replay, a combination of random errors and burst errors will occur. Data will come from tape in the sequence of the inner codewords. The 8,14 channel decoder may provide error flags to assist the inner decoder. In D-3 each inner code has 8 bytes of redundancy, so it would be possible to locate and correct up to 4 bytes in error or correct up to eight flagged errors in the codeword. This, however, is not done, because it is better to limit the correction power and use the remaining redundancy to decrease the probability of miscorrection. In this way, a smaller number of random errors are corrected with a high degree of confidence. Larger errors cause the entire inner codeword to be declared corrupt, and it is written into the appropriate row of the de-interleave array as all zeros, with error flags set. When all of the rows of the array are completed, it is possible to begin processing the columns. A burst error on tape which destroys one or more inner codewords will result in single-byte errors with flags in many different outer codewords. The presence of the error flags means that the outer code does not need to compute the position of the errors, so the full power of the outer code redundancy is available for correction. The eight redundancy bytes of D-3 can correct eight error bytes. The process of using error flags to assist a code is called erasure. It will be evident that enormous damage must be done to the tape track before four bytes are corrupted in a single outer code word, which gives high resistance to dropouts and tape scratches.

4.26 Concealment and shuffle

In the event of a gross error, correction may not be possible, and some bytes may be declared uncorrectable. In this case the errors must be concealed by interpolation from adjacent samples. Owing to the regular structure of the product code block, uncorrectable burst errors would show up as a regular pattern of interpolations on the screen. The eye is extremely good at finding such patterns, so the visibility is higher than might be expected. The solution to this problem is to perform a two-dimensional pseudo-random shuffle on the video samples before generating the product code. The reverse shuffle is performed on replay. The positions of interpolated samples due to an uncorrected error now form a pseudo random pattern with much decreased visibility.

Appendix 4.1 Calculation of Reed–Solomon generator polynomials

For a Reed–Solomon codeword over $GF(2^3)$, there will be seven 3 bit symbols. For location and correction of one symbol, there must be two redundant symbols P and Q, leaving A–E for data.

The following expressions must be true, where a is the primitive element of $x^3 \oplus x \oplus 1$ and $\oplus$ is XOR throughout:

$$A \oplus B \oplus C \oplus D \oplus E \oplus P \oplus Q = 0 \qquad (1)$$

$$a^7A \oplus a^6B \oplus a^5C \oplus a^4D \oplus a^3E \oplus a^2P \oplus aQ = 0 \qquad (2)$$

Dividing Eqn (2) by a:

$$a^6A \oplus a^5B \oplus a^4C \oplus a^3D \oplus a^2E \oplus aP \oplus Q = 0$$

$$= A \oplus B \oplus C \oplus D \oplus E \oplus P \oplus Q$$

Cancelling Q, and collecting terms:

$$(a^6 \oplus 1)A \oplus (a^5 \oplus 1)B \oplus (a^4 \oplus 1)C \oplus (a^3 \oplus 1)D \oplus (a^2 \oplus 1)E$$

$$= (a + 1)P$$

Using Section 4.20 to calculate $(a^n + 1)$, e.g. $a^6 + 1 = 101 + 001 = 100 = a^2$:

$$a^2A \oplus a^4B \oplus a^5C \oplus aD \oplus a^6E = a^3P$$

$$a^6A \oplus aB \oplus a^2C \oplus a^5D \oplus a^3E = P$$

Multiply Eqn (1)by a^2 and equating to Eqn (2):

$$a^2A \oplus a^2B \oplus a^2C \oplus a^2D \oplus a^2E \oplus a^2P \oplus a^2Q = 0$$
$$= a^7A \oplus a^6B \oplus a^5C \oplus a^4D \oplus a^3E \oplus a^2P \oplus aQ$$

Cancelling terms a^2P and collecting terms (remember $a^2 \oplus a^2 = 0$):

$$(a^7 \oplus a^2)A \oplus (a^6 \oplus a^2)B \oplus (a^5 \oplus a^2)C \oplus (a^4 \oplus a^2)D \oplus$$
$$(a^3 \oplus a^2)E = (a^2 \oplus a)Q$$

Adding powers according to Section 4.20, e.g.

$$a^7 \oplus a^2 = 001 \oplus 100 = 101 = a^6:$$
$$a^6A \oplus B \oplus a^3C \oplus aD \oplus a^5E = a^4Q$$
$$a^2A \oplus a^3B \oplus a^6C \oplus a^4D \oplus aE = Q$$

References

1. DEELEY, E.M., Integrating and differentiating channels in digital tape recording. *Radio Electron. Eng.*, **56**, 169–173 (1986)
2. MEE, C.D., *The Physics of Magnetic Recording*. Amsterdam and New York: Elsevier–North Holland (1978)
3. JACOBY, G.V., Signal equalization in digital magnetic recording. *IEEE Trans. Magn.*, **MAG-11**, 302–305 (1975)
4. SCHNEIDER, R.C., An improved pulse-slimming method for magnetic recording. *IEEE Trans. Magn.*, **MAG-11**, 1240–1241 (1975)
5. MILLER, A., U.S Patent. No. 3,108,261
6. MALLINSON, J.C. and MILLER, J.W., Optimum codes for digital magnetic recording. *Radio and Electron. Eng.*, **47**, 172–176 (1977)
7. MILLER, J.W., DC-free encoding for data transmission system. US Patent 4,027,335, (1977)
8. TANG, D.T., Run-length-limited codes. *IEEE International Symposium on Information Theory* (1969)
9. YOKOYAMA, K., Digital video tape recorder. *NHK Tech. Monogr.*, No. 31 (March 1982)
10. COLEMAN, C.H., *et al.*, High data rate magnetic recording in a single channel. *J. IERE*, **55**, 229–236 (1985)
11. KOBAYASHI, H., Application of partial response channel coding to magnetic recording systems. *IBM J. Res. Dev.*, **14**, 368–375 (1970)
12. FORNEY, G.D., jr, The Viterbi algorithm. *Proc. IEEE*, **61**, 268–278 (1973)
13. WOOD, R.W. and PETERSEN, D.A., Viterbi detection of Class IV partial response on a magnetic recording channel. *IEEE Trans. Commun.*, **34**, 454–461 (1968)
14. SHANNON, C.E., A mathematical theory of communication. *Bell Syst. Tech. J.*, **27**, 379 (1948)
15. REED, I.S. and Solomon, G., Polynomial codes over certain finite fields. *J. Soc. Ind. Appl. Math.*, **8**, 300–304 (1960)
16. BERLEKAMP, E.R., *Algebraic Coding Theory*. New York: McGraw-Hill (1967). Reprint edition: Laguna Hills, CA: Aegean Park Press (1983)
17. SUGIYAMA, Y. *et al.*, An erasures and errors decoding algorithm for Goppa codes. *IEEE Trans. Inf. Theory*, **IT-22**, (1976)
18. PETERSON, W.W. and WELDON, E.J., *Error Correcting Codes* 2nd.edn., Cambridge MA: MIT Press (1972)

Chapter 5

Digital video interfaces

Since digital video equipment is supplied by a variety of manufacturers, there is a need for a standardized interconnect so that units may communicate in the digital domain. This chapter discusses the interconnects available for conventional component and composite digital video and for 16:9 aspect ratio signals.

5.1 Introduction

Of all the advantages of digital video the most important of these for production work is the ability to pass through multiple generations without quality loss. Effects machines perform transforms on images in the digital domain which remain impossible in the analog domain. For the highest-quality post production work, digital interconnection between such items as switchers, recorders and effects machines is highly desirable to avoid the degradation due to repeated conversion and filtering stages.

In 4:2:2 digital colour difference sampling according to CCIR-601, the luminance is sampled at 13.5 MHz, which is line synchronous to both broadcast line rates, and the two colour difference signals are sampled at one-half that frequency. Composite digital machines sample at four times subcarrier. All of the signals use 8 or 10 bit resolution. Video converters universally use parallel connection, where all bits of the pixel value are applied simultaneously to separate pins. Rotary-head digital recorders lay data on the tape serially, but within the circuitry of the recorder, parallel presentation is in the majority, because it allows slower, and hence cheaper, memory chips to be used for interleaving and timebase correction. The Reed–Solomon error correction depends upon symbols assembled from several bits at once. Digital effects machines and switchers universally operate upon pixel values in parallel using fast multiplier chips.

Bearing all of this in mind, there is a strong argument for parallel connection, because the video naturally appears in the parallel format in typical machines. All that is necessary is a set of suitable driver chips, running at an appropriate sampling rate, to send video down cables having separate conductors for each bit of the sample, and clocks to tell the receiver when to sample the bit values. The cost in electronic components is very small, and for short distances this approach represents the optimum solution. An example would be where it is desired to dub a digital recording from one machine to another standing beside it.

Parallel connection has drawbacks too; these come into play when longer distances are contemplated. A multicore cable is expensive, and the connectors are physically large. It is difficult to provide good screening of a multicore cable without making it inflexible. More seriously, there are electronic problems with multicore cables. The propagation speeds of pulses down all of the cores in the cable will not be exactly the same, and so, at the end of a long cable, some bits may still be in transition when the clock arrives, while others may have begun to change to the value in the next pixel. In the presence of crosstalk between the conductors, and reflections due to suboptimal termination, the data integrity eventually becomes marginal.

Where it is proposed to interconnect a large number of units with a router, that device will be extremely complex because of the number of parallel signals to be handled.

The answer to these problems is the serial connection. All of the digital samples are multiplexed into a serial bit stream, and this is encoded to form a self-clocking channel code which can be sent down a single channel. Skew caused by differences in propagation speed cannot then occur. The bit rate necessary is in excess of 200 Mbits/s, but this is easily accommodated by coaxial cable. Provided suitable equalization and termination is used, it may be possible to send such a signal over cables intended for analog video.

The cabling savings implicit in such a system are obvious, but the electronic complexity of a serial interconnect is much greater, as high-speed multiplexers or shift registers are necessary at the transmitting end, and a phase-locked loop, data separator and deserializer are needed at the receiver to regenerate the parallel signal needed within the equipment. Although this increased complexity raises cost, it has to be offset against the saving in cabling. The availability of specialized chips is speeding the acceptance of serial transmission.

A distinct advantage of serial transmission is that a matrix distribution unit or router is more easily realized. Where numerous pieces of video equipment need to be interconnected in various ways for different purposes, a crosspoint matrix is an obvious solution. With serial signals, only one switching element per signal is needed, whereas in a parallel system, a matrix would be unwieldy. A serial system has a potential disadvantage that the time distribution of bits within the block has to be closely defined, and, once standardized, it is extremely difficult to increase the wordlength if this is found to be necessary. The CCIR 8 bit serial system suffered from this problem, and has been replaced by a new serial standard which incorporates two extension bits which can be transmitted as zero in 8-bit applications, but allows 10-bit use if necessary. In a parallel interconnect, the word extension can be achieved by adding extra conductors alongside the existing bits, which is much easier.

The third interconnect to be considered uses fibre optics. The advantages of this technology are numerous. The bandwidth of an optical fibre is staggering, as it is determined primarily by the response speed of the light source and sensor. The optical transmission is immune to electromagnetic interference from other sources, nor does it contribute any. This is advantageous for connections between cameras and control units, where a long cable run may be required in outside broadcast applications. The cable can be made completely from insulating materials, so that ground loops cannot occur, although many practical fibre-optic cables include electrical conductors for power and steel strands for mechanical strength.

Drawbacks of fibre optics are few. They do not like too many connectors in a given channel, as the losses at a connection are much greater than with an electrical plug and socket. It is preferable for the only breaks in the fibre to be at the transmitting and receiving points. For similar reasons, fibre optics are less suitable for distribution, where one source feeds many destinations. The familiar loop-through connection of analog video is just not possible. The bidirectional open-collector or tri-state buses of electronic systems cannot be implemented with fibre optics, nor is it easy to build a crosspoint matrix.

5.2 Digital video interfaces

At the time of writing standards or advanced proposals exist for both 4:2:2 component and $4F_{sc}$ composite digital in both 525/59.94 and 625/50 and in both parallel and serial forms. Interfaces also exist for widescreen applications. All digital interfaces require to be standardized in the following areas: connectors, to ensure plugs mate with sockets; pinouts; electrical signal specification, to ensure that the correct voltages and timing are transferred; and protocol, to ensure that the meaning of the data words conveyed is the same to both devices. As digital video of any type is only data, it follows that the same physical and electrical standards can be used for a variety of protocols.

The parallel interface uses common connectors, pinouts and electrical levels for both line standards in both the component and composite versions. The same is true for the serial interfaces. Thus there are only two electrical interfaces: parallel and serial. The type of video being transferred is taken care of in the protocol differences.

5.3 The parallel electrical interface

Composite signals use the same electrical and mechanical interface as is used for 4:2:2 component working.[1,2] This means that it is possible to plug erroneously a component signal into a composite machine. Whilst this cannot possibly work, no harm wll be done because the signal levels and pinouts are the same.

Each signal in the interface is carried by a balanced pair using ECL drive levels with a nominal impedance of 110 ohms. As Figure 5.1 shows, there are eight signal pairs and two optional pairs, so that a 10 bit word can be accommodated. The optional signals are used to add bits at the least significant end of the word. Adding bits in this way extends resolution rather than increasing the magnitude. It will be seen from the figure that the optional bits are called Data −1 and Data −2 where the −1 and −2 refer to the power of 2 represented, i.e. 2^{-1} and 2^{-2}. The 8 bit word ends in a radix point and the extra bits below the radix point represent the half and quarter quantizing intervals. In this way a degree of compatibility exists between 10 and 8 bit systems, as the correct magnitude will always be obtained when changing wordlength, and all that is lost is a degree of resolution in shortening the wordlength when the bits below the radix point are lost. The same numbering scheme can be used for both wordlengths; the longer wordlength simply has a radix point and an extra digit in any number base. Converting to the 8 bit equivalent is then simply a matter of deleting the extra digit and retaining the integer part.

A separate clock signal pair and a number of grounding and shielding pins complete the connection. A 25 pin D-type connector to ISO 2110–1989 is

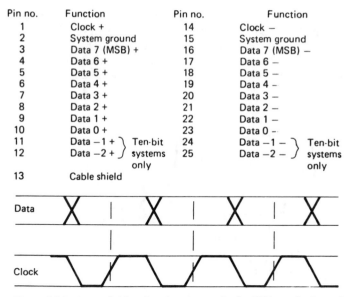

Pin no.	Function	Pin no.	Function
1	Clock +	14	Clock −
2	System ground	15	System ground
3	Data 7 (MSB) +	16	Data 7 (MSB) −
4	Data 6 +	17	Data 6 −
5	Data 5 +	18	Data 5 −
6	Data 4 +	19	Data 4 −
7	Data 3 +	20	Data 3 −
8	Data 2 +	21	Data 2 −
9	Data 1 +	22	Data 1 −
10	Data 0 +	23	Data 0 −
11	Data −1 + ⎫ Ten-bit	24	Data −1 − ⎫ Ten-bit
12	Data −2 + ⎭ systems only	25	Data −2 − ⎭ systems only
13	Cable shield		

Figure 5.1 In the parallel interface there is capacity for 10 bit words plus a clock where each signal is differential. In 8 bit systems 'Data −1' and 'Data −2' are not used. Clock transition takes place in the centre of a data bit cell as shown below.

specified. Equipment always has female connectors, cables always have male connectors. Metal or metallized backshells are recommended for optimum shielding. Whilst equipment may produce or accept only 8 bit data, cables must be wired for all 10 bits.

Connector latching is by a pair of 4–40 (an American thread) screws, with suitable posts provided on the female connector. It is important that the screws are used as the multicore cable is quite stiff and can eventually unseat the plug if it is not secured. Some early equipment had slidelocks instead of screw pillars, but these proved to be too flimsy.

Figure 5.1 also shows the relationship between the clock and the data. A positive-going clock edge is used to sample the signal lines after the level has settled between transitions. In 4:2:2, the clock will be line-locked 27 MHz irrespective of the line standard, whereas in composite digital the clock will be four times the frequency of PAL or NTSC subcarrier.

The parallel interface is suitable for distances of up to 50 metres. Beyond this distance equalization is likely to be necessary and skew or differential delay between signals may become a problem. For longer distances a serial interface is a better option.

5.4 The 4:2:2 parallel interface

It is not necessary to digitize sync in component systems, since the sampling rate is derived from sync. The only useful video data are those sampled during the active line. All other parts of the video waveform can be recreated at a later time. It is only necessary to standardize the size and position of a digital active line.

The position is specified as a given number of sampling-clock periods from the leading edge of sync, and the length is simply a standard number of samples. The component digital active line is 720 luminance samples long. This is slightly longer than the analog active line and allows for some drift in the analog input. Ideally the first and last samples of the digital active line should be at blanking level.

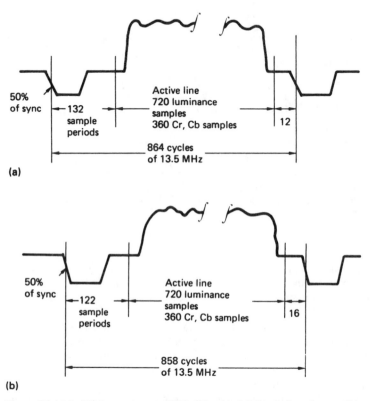

Figure 5.2 (a) In 625 line systems to CCIR-601, with 4:2:2 sampling, the sampling rate is exactly 864 times line rate, but only the active line is sampled, 132 sample periods after sync. (b) In 525 line systems to CCIR-601, with 4:2:2 sampling, the sampling rate is exactly 858 times line rate, but only the active line is sampled, 122 sample periods after sync. Note active line contains exactly the same quantity of data as for 50 Hz systems.

Figure 5.2 shows that in 625 line systems[3] the control system waits for 132 sample periods before commencing sampling the line. Then 720 luminance samples and 360 of each type of colour difference sample are taken, 1440 samples in all. A further 12 sample periods will elapse before the next sync edge, making 132 + 720 + 12 = 864 sample periods. In 525 line systems,[2] the analog active line is in a slightly different place and so the controller waits 122 sample periods before taking the same digital active line samples as before. There will then be 16 sample periods before the next sync edge, making 122 + 720 + 16 = 858 sample periods.

Figure 5.3 shows the luminance signal sampled at 13.5 MHz and two colour difference signals sampled at 6.75 MHz. Three separate signals with different clock rates are inconvenient and so multiplexing can be used. If the colour difference signals are multiplexed into one channel, then two 13.5 MHz channels will be required. If these channels are multiplexed into one, a 27 MHz clock will be required. The word order will be:

C_b, Y, C_r, Y, etc.

In order unambiguously to deserialize the samples, the first sample in the line is always C_b.

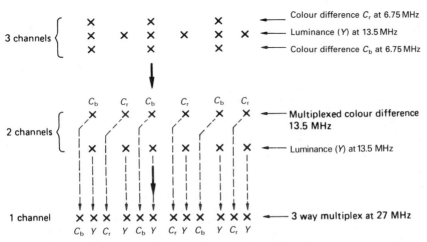

Figure 5.3 The colour difference sampling rate is one-half that of luminance, but there are *two* colour difference signals, C_r and C_b, hence the colour difference data rate is equal to the luminance data rate and a 27 MHz interleaved format is possible in a single channel.

In addition to specifying the location of the samples, it is also necessary to standardize the relationship between the absolute analog voltage of the waveform and the digital code value used to express it so that all machines will interpret the numerical data in the same way. These relationships are in the voltage domain and are independent of the line standard used.

Figure 5.4 shows how the luminance signal fits into the quantizing range of an 8 bit system. Black is at a level of 16_{10} and peak white is at 235_{10} so that there is some tolerance of imperfect analog signals. The sync pulse will clearly go outside the quantizing range, but this is of no consequence as conventional syncs are not transmitted. The visible voltage range fills the quantizing range and this gives the best possible resolution.

The colour difference signals use offset binary, where 128_{10} is the equivalent of blanking voltage. The peak analog limits are reached at 16_{10} and 240_{10} respectively, allowing once more some latitude for maladjusted analog inputs.

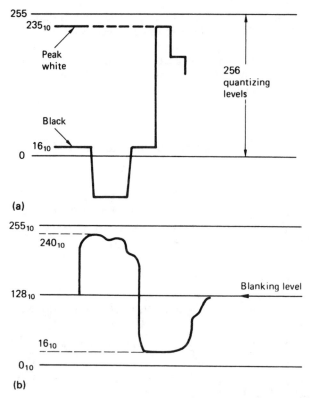

Figure 5.4 (a) The luminance signal fits into the quantizing range with a little allowance for excessive gain. Here black is at 16_{10} and peak white is at 235_{10}. The sync pulse goes outside the quantizing range but this is of no consequence as it is not transmitted. (b) The colour difference signals use offset binary where blanking level is at 128_{10}, and the peaks occur at 16_{10} and 240_{10} respectively.

Note that the code values corresponding to all ones and all zeros, i.e. the two extreme ends of the quantizing range, are not allowed to occur in the active line as they are reserved for synchronizing. Converters must be followed by circuitry which catches these values and forces the LSB to a different value if out-of-range analog inputs are applied.

The peak-to-peak amplitude of Y is 220 quantizing intervals, whereas for the colour difference signals it is 225 intervals. There is thus a small gain difference between the signals. This will be cancelled out by the opposing gain difference at any future DAC, but must be borne in mind when digitally converting to other standards.

As conventional syncs are not sent, horizontal and vertical synchronizing is achieved by special bit patterns sent with each line. Immediately before the digital active line location is the *SAV* (start of active video) pattern, and immediately after is the *EAV* (end of active video) pattern. These unique patterns occur on every line and continue throughout the vertical interval.

Each sync pattern consists of four symbols. The first is all ones and the next two are all zeros. As these cannot occur in active video, their detection reliably indicates a sync pattern.

The fourth symbol is a data byte which contains three data bits, H, F and V. These bits are protected by four redundancy bits which form a 7-bit Hamming codeword for the purpose of detecting and correcting errors.

Figure 5.5(a) shows the structure of the sync pattern. The sync bits have the following meanings:

H is used to distinguish between SAV, where it is set to 0 and EAV where it is set to 1.

F defines the state of interlace and is 0 during the first field and 1 during the second field. F is only allowed to change at EAV. In interlaced systems, one field begins at the centre of a line, but there is no sync pattern at that location so the field bit changes at the end of the line in which the change took place.

V is 1 during vertical blanking and 0 during the active part of the field. It can only change at EAV.

Figure 5.5(b) (top) shows the relationship between the sync pattern bits and 625 line analog timing, whilst below is the relationship for 525 lines. Only the active line is transmitted and this leaves a good deal of spare capacity. The two line standards differ on how this capacity is used. In 625 lines, only the active line period may be used on lines 20 to 22 and 333 to 335.[3] Lines 20 and 333 are reserved for equipment self-testing.

In 525 lines there is considerably more freedom and ancillary data may be inserted anywhere there is no active video, either during horizontal blanking, vertical blanking, or both.[2]

The all-zeros and all-ones codes are reserved for synchronizing, and cannot be allowed to appear in ancillary data. In practice only 7 bits of the 8 bit word can be used as data; the eighth bit is redundant and gives the byte odd parity. As all ones and all zeros are even parity, the sync pattern cannot then be generated accidentally.

For 16:9 aspect ratio working, the line and field rate remain the same, but the luminance sampling rate may be raised to 18 Mhz and the colour difference sampling rates are raised to 9 MHz. This results in the sampling structure shown

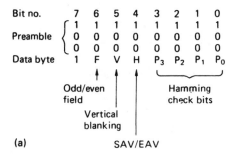

(a)   SAV/EAV

Figure 5.5(a) The 4 byte synchronizing pattern which precedes and follows every active line sample block has this structure.

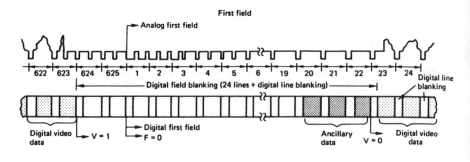

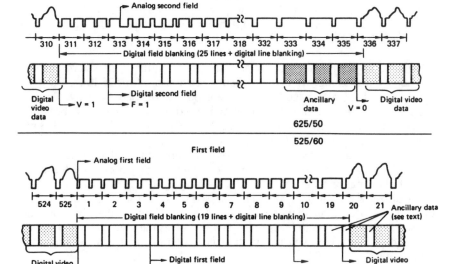

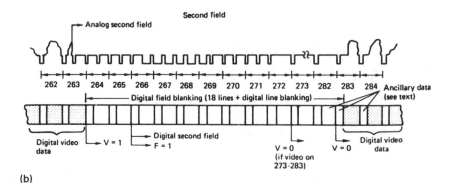

(b)

Figure 5.5(b) The relationships between analog video timing and the information in the digital timing reference signals for 625/50 (above) and 525/60 (below).

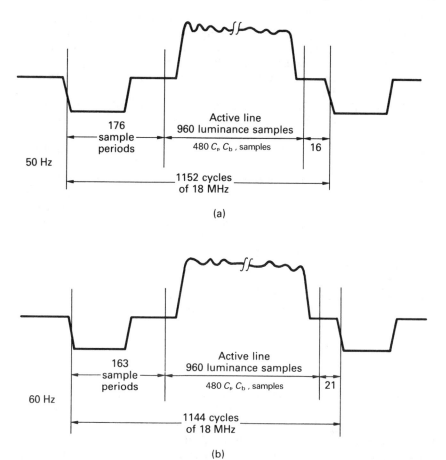

Figure 5.6 In 16:9 working with an 18 MHz sampling rate the sampling structure shown here results.

in Figure 5.6. There are now 960 luminance pixels and 2 × 480 colour difference pixels. The parallel interface remains the same except that the clock rate rises to 36 MHz.

5.5 The composite digital parallel interface

Composite digital samples at four times subcarrier frequency, and so there will be major differences between the standards.

Whilst the 4:2:2 interface transmits only active lines and special sync patterns, the composite interfaces carry the entire composite waveform: syncs, burst and all. Although ancillary data may be placed in sync tip, the sync edges must be present. In the absence of ancillary data, the data on the parallel interface are essentially the continuous stream of samples from a converter which is digitizing a normal analog composite signal. Virtually all that is necessary to return to the analog domain is a DAC and a filter. One of the reasons for this different approach is that the sampling clock in composite video is subcarrier based. The sample values during sync can change with ScH phase in NTSC and PAL and change with the position in the frame in PAL due to the 25 Hz component. It is simpler to convey sync sample values on the interface than to go to the trouble of re-creating them later.

The instantaneous voltage of composite video can go below blanking on dark saturated colours, and above peak white on bright colours. As a result the quantizing ranges need to be stretched in comparison with 4:2:2 in order to accommodate all possible voltage excursions. Sync tip can be accommodated at the low end and peak white is some way below the end of the scale. It is not so easy to determine when overload clipping will take place in composite as the sample sites are locked to subcarrier. The degree of clipping depends on the chroma phase. When samples are taken either side of a chroma peak, clipping will be less likely to occur than when the sample is taken at the peak. Advantage is taken of this phenomenon in PAL as the peak analog voltage of a 100% yellow bar goes outside the quantizing range. The sampling phase is such that samples are sited either side of the chroma peak and remain within the range.

The PAL and NTSC versions of the composite digital interface will be described separately. The electrical interface is the same for both, and was described in Section 5.3.

5.6 PAL interface

The quantizing range of digital PAL is shown in Figure 5.7.[4] Blanking level is at 64_{10} or 40 hex and sync tip is the lowest allowable code of 1 as 0 is reserved for digital synchronizing along with 255. Peak white is 211_{10} or D3 hex. As with 4:2:2, two optional bits can be added below the LSB to extend the resolution. In 10 bit working blanking becomes 40.0 hex and peak white becomes D3.0 hex, although some documents will be found which confusingly shift the radix point down in 10 bit mode and denote blanking by 100 Hex and peak white by 34C hex.

In PAL, the composite digital interface samples at $4 \times F_{sc}$ with sample phase aligned with burst phase. PAL burst swing results in burst phases of ±135 degrees, and samples are taken at these phases and at ±45 degrees, precisely half-way between the U and V axes. This sampling phase is easy to generate from burst and avoids premature clipping of chroma. It is most important that samples are taken exactly at the points specified, since any residual phase error in the sampling clock will cause the equivalent of a chroma phase error when samples from one source are added to samples from a different source in a switcher. A digital switcher can only add together pairs of samples from different inputs, but if these samples were not taken at the same instants with respect to their subcarriers, the samples represent different vectors and cannot be added.

Figure 5.8 shows how the sampling clock may be derived. The incoming sync is used to derive a burst gate, during which the samples of burst are analysed. If the clock is correctly phased, the sampled burst will give values of 95_{10}, 64_{10}, 32_{10}, 64_{10}, repeated, whereas if a phase error exists, the values at the burst

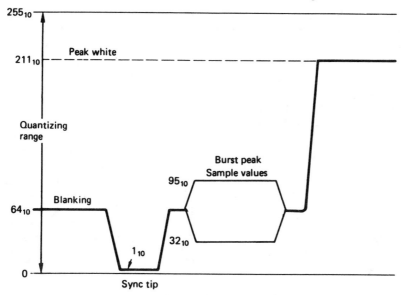

Figure 5.7 The composite PAL signal fits into the quantizing range as shown here. Note that there is sufficient range to allow the instantaneous voltage to exceed that of peak white in the presence of saturated bright colours. Values shown are decimal equivalents in a 10 or 8 bit system. In a 10 bit system the additional 2 bits increase resolution, not magnitude, so they are below the radix point and the decimal equivalent is unchanged. PAL samples in phase with burst, so that values shown are on the burst peaks and are thus also the values of the envelope.

crossings will be above or below 64_{10}. The difference between the sample values and blanking level can be used to drive a DAC which controls the sampling VCO. In this way any phase errors in the ADC are eliminated, because the sampling clock will automatically servo its phase to be identical to digital burst. Burst swing causes the burst peak and burst crossing samples to change places, so a phase comparison is always possible during burst. DC level shifts can be removed by using both positive and negative burst crossings and averaging the results. This also has the effect of reducing the effect of noise.

In PAL, the subcarrier frequency contains a 25 Hz offset, and so $4 \times F_{sc}$ will contain a 100 Hz offset. The sampling rate is not h-coherent, and the sampling structure is not quite orthogonal. As subcarrier is given by:

$$F_{sc} = 283\tfrac{3}{4}F_h + F_v/2$$

The sampling rate will be given by:

$$F_s = 1135F_h + 2F_v$$

This results in 709 379 samples per frame, and there will not be a whole number of samples in a line. In practice, 1135 sample periods, numbered 0 to 1134, are defined as one digital line, with an additional four sample periods per frame which are included by having 1137 samples, numbered 0 to 1136, in

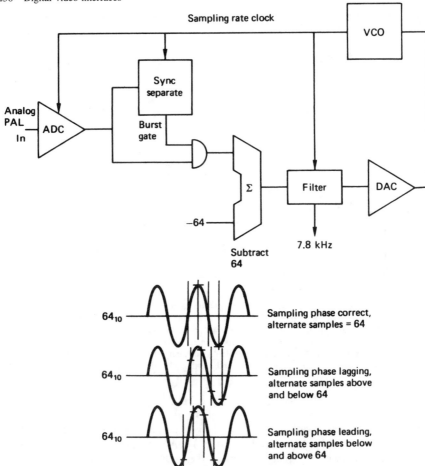

Figure 5.8 Obtaining the sample clock in PAL. The values obtained by sampling burst are analysed. When phase is correct, burst will be sampled at zero crossing and sample value will be 64_{10} or blanking level. If phase is wrong, sample will be above or below blanking. Filter must ignore alternate samples at burst peaks and shift one sample every line to allow for burst swing. It also averages over several burst crossings to reduce jitter. Filter output drives DAC and thus controls sampling clock VCO.

lines 313 and 625. Figure 5.9(a) shows the sample numbering scheme for an entire line. Note that the sample numbering begins at 0 at the start of the digital active line so that the horizontal blanking area is near the end of the digital line and the sample numbers will be large. The digital active line is 948 samples long and is longer than the analog active line. This allows the digital active line to move with 25 Hz whilst ensuring the entire analog active line is still conveyed.

Since sampling is not h-coherent, the position of sync pulses will change relative to the sampling points from line to line. The relationship can also be changed by the ScH phase of the analog input. Zero ScH is defined as coincidence between sync and zero degrees of subcarrier phase at line 1 of

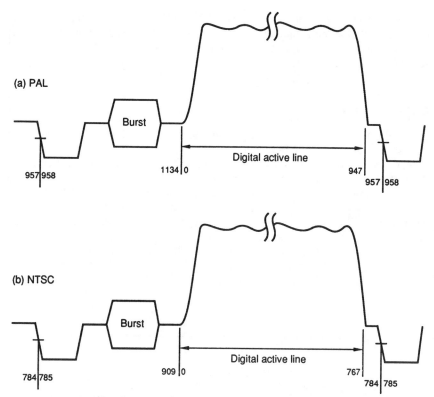

Figure 5.9 (a) Sample numbering in digital PAL. There are defined to be 1135 sample periods per line of which 948 are the digital active line. This is longer than the analog active line. Two lines per frame have two extra samples to compensate for the 25 Hz offset in subcarrier. NTSC is shown in (b). Here there are 910 samples per line of which 768 are the digital active line.

field 1. Since composite digital samples on burst phase, not on subcarrier phase, the definition of zero ScH will be as shown in Figure 5.10, where it will be seen that two samples occur at exactly equal distances either side of the 50% sync point. If the input is not zero ScH, the samples conveying sync will have different

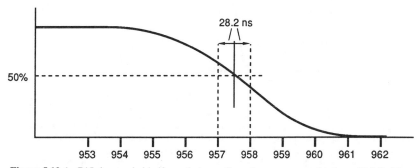

Figure 5.10 As PAL is sampled half-way between the colour axes, sample sites will fall either side of 50% sync at the zero ScH measurement line.

values. Measurement of these values will allow ScH phase to be computed. In a DVTR installation, non-standard ScH is only a problem on the initial conversion from analog because composite DVTRs do not record sync, and will regenerate zero ScH syncs on replay.

5.7 NTSC interface

Although they have some similarities, PAL and NTSC are quite different when analysed at the digital sample level.

Figure 5.11 shows how the NTSC waveform fits into the quantizing structure.[5] Blanking is at 60_{10} in an 8 bit system, and peak white is at 200_{10}, so that 1 IRE unit is the equivalent of $1.4Q$ which could perhaps be called the DIRE. These different values are due to the different sync/vision ratio of NTSC. PAL is 7:3 whereas NTSC is 10:4. As with 4:2:2 and PAL, two optional extra bits at the LSB end of the sample can extend the resolution.

Subcarrier in NTSC has an exact half-line offset, so there will be an integer number of cycles of subcarrier in two lines. F_{sc} is simply $227.5 \times F_h$, and as sampling is at $4 \times F_{sc}$, there will be $227.5 \times 4 = 910$ samples per line period,

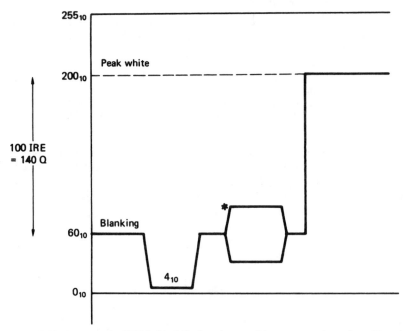

Figure 5.11 The composite NTSC signal fits into the quantizing range as shown here. Note that there is sufficient range to allow the instantaneous voltage to exceed peak white in the presence of saturated, bright colours. Values shown are decimal equivalents in an 8 or 10 bit system. In a 10 bit system the additional 2 bits increase resolution, not magnitude, so they are below the radix point and the decimal equivalent is unchanged. *Note that, unlike PAL, NTSC does not sample on burst phase and so values during burst are not shown here. See Figure 5.12 for burst sample details.

and the sampling will be orthogonal. Figure 5.9(b) shows that the digital active line consists of 768 samples numbered 0 to 767. Horizontal blanking follows the digital active line in sample numbers 768 to 909.

The sampling phase is chosen to facilitate encoding and decoding in the digital domain. In NTSC there is a phase shift of 123 degrees between subcarrier and the I axis. As burst is an inverted piece of the subcarrier waveform, there is a phase shift of 57 degrees between burst and the I axis. Composite digital NTSC does not sample in phase with burst, but on the I and Q axes at 57, 147, 237 and 327 degrees with respect to burst.

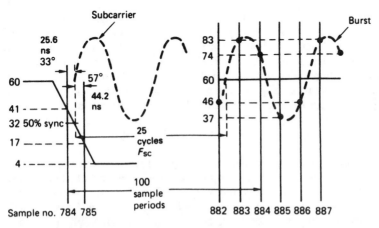

Figure 5.12 NTSC ScH phase. Sampling is not performed in phase with burst as in PAL, but on the I and Q axes. Since in NTSC there is a phase angle of 57° between burst and I, this will also be the phase at which burst samples should be taken. If ScH phase is zero, then phase of subcarrier taken at 50% sync will be zero, and the samples will be taken 33° before and 57° after sync; 25 cycles of subcarrier or 100 samples later, during burst, the sample values will be obtained. Note that in NTSC burst is inverted subcarrier, so sample 785 is positive, but sample 885 is negative.

Figure 5.12 shows how this approach works in relation to sync and burst. Zero ScH is defined as zero degrees of subcarrier at the 50% point on sync, but the 57 degree sampling phase means that the sync edge is actually sampled 25.6 ns ahead of, and 44.2 ns after the 50% point. Similarly, when the burst is reached, the phase shift means that burst sample values will be 46_{10}, 83_{10}, 74_{10} and 37_{10} repeating. The phase-locked loop which produces the sampling clock will digitally compare the samples of burst with the values given here. Since it is not sampling burst at a zero crossing, the slope will be slightly less, so the gain of the phase error detector will also be less, and more prone to burst noise than in the PAL process. The phase error can, however, be averaged over several burst samples to overcome this problem.

As in PAL, if the analog input does not have zero ScH phase, the sync pulse values will change, but burst values will not. As in PAL, NTSC DVTRs do not record sync, and will regenerate zero ScH syncs on replay.

5.8 Serial digital interface (SDI)

The interface described here has been developed to allow up to 10 bit samples of component or composite digital video to be communicated serially.[6] 16:9 format component signals with 18 MHz sampling rate can also be handled. The interface allows ancillary data including transparent conveyance of AES/EBU digital audio channels.

Scrambling, or pseudo-random coding, uses concepts which were introduced in Chapter 4. The serial interface uses convolutional coding, which is simpler to implement in a cable installation because no separate synchronizing of the randomizing is needed.

The components necessary for a composite serial link are shown in Figure 5.13. Parallel component or composite data having a wordlength of up to 10 bits form the input. These are fed to a 10 bit shift register which is clocked at ten times the input rate, which will be 270 MHz or $40 \times F_{sc}$. If there are only 8 bits in the input words, the missing bits are forced to zero for transmission except for the all-ones condition which will be forced to ten ones. The serial data from the shift register are then passed through the scrambler, in which a given bit is converted to the exclusive OR of itself and 2 bits which are five and nine clocks ahead. This is followed by another stage, which converts channel ones into transitions. The resulting signal can be fed down 75 Ω coaxial cable using BNC connectors.

The scrambling process at the transmitter spreads the signal spectrum and makes that spectrum reasonably constant and independent of the picture content. It is possible to assess the degree of equalization necessary by comparing the energy in a low-frequency band with that in higher frequencies. The greater the disparity, the more equalization is needed. Thus fully automatic cable equalization is easily achieved. The receiver must generate a bit clock at 270 MHz or $40 \times F_{sc}$ from the input signal, and this clock drives the input sampler and slicer which converts the cable waveform back to serial binary. The local bit clock also drives a circuit which simply reverses the scrambling at the transmitter. The first stage returns transitions to ones, and the second stage is a mirror image of the encoder which reverses the exclusive OR calculation to output the original data. Since transmission is serial, it is necessary to obtain word synchronization, so that correct deserialization can take place.

In the component parallel input, the *SAV* and *EAV* sync patterns are present and the all-ones and all-zeros bit patterns these contain can be detected in the shift register and used to reset the deserializer.

In the composite parallel interface signal, there are no equivalents of the 4:2:2 sync patterns and it is necessary to create an equivalent. The timing reference and identification signal (TRS-ID) is added during blanking, and the receiver can detect the patterns which it contains. TRS-ID consists of five words which are inserted just after the leading edge of video sync. Figure 5.14(a) shows the location of TRS-ID at samples 967–971 in PAL and Figure 5.14(b) shows the location at samples 790–794 in NTSC.

Out of the five words in TRS-ID, the first four are for synchronizing, and consist of a single word of all ones, followed by three words of all zeros. The fifth word is for identification, and carries the line and field numbering information shown in Figure 5.15. The field numbering is colour framing

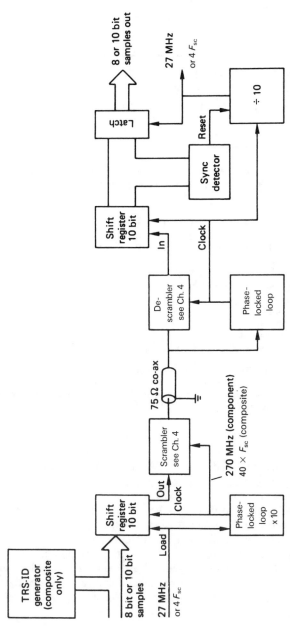

Figure 5.13 Major components of a serial scrambled link. Input samples are converted to serial form in a shift register clocked at ten times the sample rate. The serial data are then scrambled for transmission. On reception, a phase-locked loop recreates the bit rate clock and drives the de-scrambler and serial-to-parallel conversion. On detection of the sync pattern, the divide-by-ten counter is rephased to load parallel samples correctly into the latch. For composite working the bit rate will be 40 times subcarrier, and a sync pattern generator (top left) is needed to inject TRS-ID into the composite data stream. See Figure 5.15 for TRS-ID detail.

(a) PAL

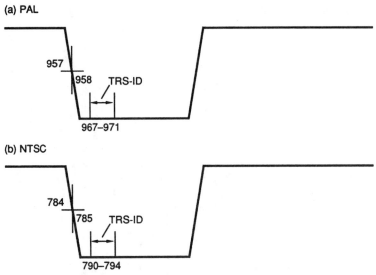

(b) NTSC

Figure 5.14 In composite digital it is necessary to insert a sync pattern during analog sync tip to ensure correct deserialization. The location of TRS-ID is shown in (a) for PAL and in (b) for NTSC.

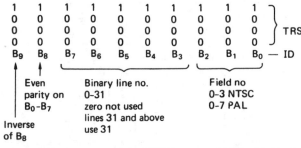

Figure 5.15 The contents of the TRS-ID pattern which is added to the transmission during the horizontal sync pulse just after the leading edge. The field number conveys the composite colour framing field count, and the line number carries a restricted line count intended to give vertical positioning information during the vertical interval. This count saturates at 31 for lines of that number and above.

information which is useful for editing. In PAL the field numbering will go from zero to seven, whereas in NTSC it will only reach three.

On detection of the synchronizing symbols, a divide-by-ten circuit is reset, and the output of this will clock words out of the shift register at the correct times. This output will also become the output word clock.

It is a characteristic of all randomizing techniques that certain data patterns will interact badly with the randomizing algorithm to produce a channel waveform which is low in clock content. These so-called pathological data patterns[7] are extremely rare in real program material, but can be generated in test equipment and used to check the adjustment of the phase-locked loop in a receiver.

5.9 Serial digital routing

Digital routers have the advantage that they need cause no loss of signal quality as they simply pass on a series of numbers. Analog routers inevitably suffer from crosstalk and noise however well made, and this reduces signal quality on every pass.

Routers are available for serial and parallel formats. The parallel video output signal is cheaper to produce as most equipment has parallel format internally. On the other hand a parallel router is potentially complex because of the large number of conductors which need switching for each input.

Serial digital signals are more expensive to output as extra circuitry is required in each device. On the other hand, a serial router is potentially very inexpensive as it is a single-pole device. It can be easier to build than an analog router because the digital signal is more resistant to crosstalk.

As a result, parallel routers are more economical for small installations, whereas the cost rapidly escalates as the router size increases. For large routers, the only logical approach is to use serial, as the economy to be had in the router offsets the requirement for serializers on all contributing equipment.

A digital router can be made using analog switches, so that the input waveform is passed from input to output. This is not the best approach, as the total length of cable which can be used is restricted as the input cable is effectively in series with the output cable and the analog losses in both will add.

A better approach is for each router input to reclock and slice the waveform back to binary. The router itself then routes logic levels and relaunches a clean signal. The cables to and from the router can be maximum length as the router is also a repeater.

It is not necessary to unscramble the serial signal at the router. A phase-locked loop is used to regenerate the bit clock. This rejects jitter on the incoming waveform. The waveform is sliced, and the slicer output is sampled by the local clock. The result is a clean binary waveform, identical to the original driver waveform.

The router is simply a binary bit stream switch, and is not unduly concerned with the meaning or content of the bit stream. It does not matter whether the bitstream is PAL, NTSC or 4:2:2 or whether or not ancillary data are carried: it just passes through.

The only parameter of any consequence is the bit rate. 4:2:2 runs at 270 Mbits/s, PAL runs at 177 Mbits/s and NTSC runs at 143 Mbits/s. 16:9 component video with 18 MHz luminance sampling rate results in 360 Mbits/s.

Some routers require a link or DIP switch to be set in each input in order to centre the VCO at the correct frequency. Otherwise they are standards independent, which allows more flexibility and economy. With a mixed-standard router, it is only necessary to constrain the control software so that inputs of a given standard can only be routed to outputs connected to devices of the same standard, and one router can then handle component and composite signals simultaneously.

5.10 Timing in digital installations

The issue of signal timing has always been critical in analog video, but the adoption of digital routing relaxes the requirements considerably. Analog vision mixers need to be fed by equal-length cables from the router to prevent propagation delay variation. In the digital domain this is no longer an issue as delay is easily obtained and each input of a digital vision mixer can have its own local timebase corrector. Provided signals are received having timing within the window of the inputs, all inputs are retimed to the same phase within the mixer.

Figure 5.16 shows how a mixing suite can be timed to a large SDI router. Signals to the router are phased so that the router output is aligned to station reference within a microsecond or so. The delay in the router may vary with its configuration but only by a few microseconds. The mixer reference is set with respect to station reference so that local signals arrive towards the beginning of the input windows and signals from the router (which, having come further, will be the latest) arrive towards the end of the windows. Thus all sources can be retimed within the mixer and any signal can be mixed with any other. Clearly the mixer introduces delay, and the signal feed back to the router experiences further delay. In order to send the mix back to the router a frame synchronizer is needed on the output of the suite. This introduces somewhat less than a frame of delay so that by the time the signal has re-emerged from the router it is aligned to station reference once more, but a frame late.

In an ideal world, every piece of hardware in the station will have component SDI outputs and inputs, and everything is connected by SDI cables. In practice,

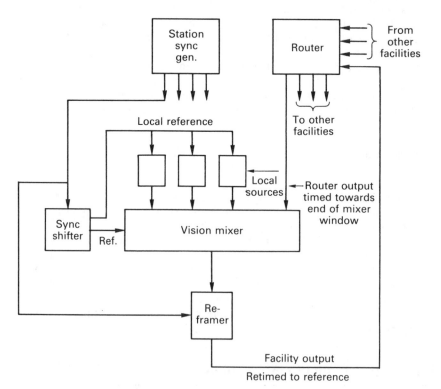

Figure 5.16 In large systems some care is necessary with signal timing as is shown here. See text for details.

unless the building is new, this is unlikely. However, there are ways in which SDI can be phased in alongside analog systems. An expensive way of doing this is to fit every composite device with coders and decoders, and every analog device with converters so that a component SDI router can be used alone. Figure 5.17 shows an alternative. The SDI router is connected to every piece of digital equipment, using parallel to serial adaptors where necessary. Analog VTRs such as Betacam SP or M-II have digital output options which use the data from the TBC directly. One thing to check with digital output SP and M-II machines is that the Y/C timing must be adjusted correctly at the TBC input in the VTR as it cannot be corrected subsequently. These will need DACs to convert SDI signals to analog component inputs. Analog equipment continues to be connected to an analog router, and interconnection paths, called gateways, are created between the two routers. These gateways require converters in each direction, but the number of converters is much less than if every analog device was equipped. The number of gateways will be determined by the number of simultaneous transactions between analog and digital domains.

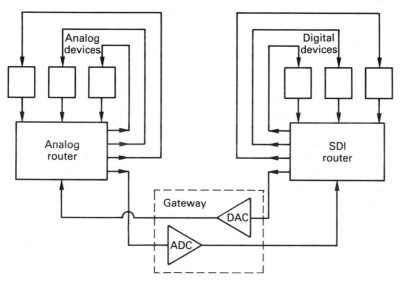

Figure 5.17 Digital routers can operate alongside existing analog routers using gateways as an economical means of introducing digital routing.

In many cases the two routers can be made to appear like one large router if appropriate software is available in a common control system. This will only be possible if the SDI router is purchased from the same manufacturer as the existing analog router or if the SDI router manufacturer offers custom software.

Whilst modern DVTRs incorporate the audio data in the SDI signal, adapting older devices to do this, and subsequently demultiplexing the audio, could prove expensive. It may be more cost effective to retain an earlier audio router, or to have a separate AES/EBU digital audio router layer controlled in parallel with the SDI router.

If the analog router handles composite signals, then coders and decoders will be needed in addition to the converters. In this case the use of gateways obviates unnecessary codecs when composite devices are connected together.

5.11 Testing digital video interfaces

Upon installing a new interface it is necessary to test it for data integrity. It is not adequate to see if the picture looks OK. The very nature of digital transmission gives beneficial resistance to minor noises and cable loss with no effect on the data and thus no visible picture artefacts. The down side of this characteristic is that a digital system can be on the verge of failure and there are no visible symptoms. If one sees a perfect picture on a monitor, it is not possible to say whether the data which produced it was received with a good noise margin which will resist future degradations or whether it was only just received and is likely to fail in the near future. Some more objective testing means is necessary if any confidence is to be had in the long-term reliability of an installation.

It is relatively simple to assess the eye pattern generated by the received waveform. After equalization the eye opening should be clearly visible, and the size of the opening should be consistent with the length and type of cable used. If it is not, the noise and/or jitter margin may be inadequate. Testing the eye pattern on the scrambled serial interface requires a fast oscilloscope, or even a sampling scope. The scope probe cannot be just hooked on to the cable as the impedance mismatch this causes will have a damaging effect on the signal. Equally a coaxial T-piece is ruled out.

One potential problem area which is frequently overlooked is to ensure that the VCO in the receiving phase-locked loop is correctly centred. If it is not, it will be running with a static phase error and will not sample the received waveform at the centre of the eyes. The sampled bits will be more prone to noise and jitter errors. VCO centring can simply be checked with a digital frequency meter which should display the correct frequency when the input is momentarily disconnected.

Inspecting the eye pattern is a good technique for establishing that the basic installation is sound and has proper signal levels, termination impedance and equalization, but is not very good at detecting infrequent impulsive noise. Contact noise from electrical power installations such as air conditioners is unlikely to be present for long enough to find with an eye pattern display. The technique of signature analysis is better suited to impulsive noise problems.

Signature analysis requires a stationary test signal to be applied at the transmitting end of the interface. In this context stationary means that the same data are repeated in each frame or, in the case of composite, in each colour frame period. The received data are then processed to generate a value known as a signature. If the transmitted data are always the same, any change in the received signature indicates that an error has occurred.

The signature generation process expresses an entire frame in component or an entire colour sequence in composite as a single word. A change of a single bit is enough to change the signature. A number of bit errors up to the number of bits in the word will guarantee to change the signature, whereas larger numbers of errors are detected with a high degree of probability.

The signature having these characteristics can be obtained by dividing the received data by a polynomial as was explained in Chapter 4. The remainder after the division is the signature. In error-detection systems, the data to be transmitted are converted to a codeword which gives a remainder or signature of zero when divided by the generating polynomial. Any non-zero remainder, known as a syndrome, indicates an error. Signature analysis is similar except that the transmitted data are not turned into a codeword. This avoids complexity at the transmitting end and the need to reserve word positions in the data. As the transmitted data are not a codeword, the remainder will be non-zero. This does not, however, indicate an error, but is a characteristic or signature of the data. An error is indicated when the signature changes, not by its absolute value.

Signature analysis is a relative detection method, rather than an absolute method. It cannot detect permanent errors, such as a stuck bit in a parallel interface, as the same signature will always be obtained. However, if the signature of the test pattern in use is known, it will be seen that the received signature is permanently different. The test pattern signature can easily be obtained by connecting the generator directly to the signature analyser as well as to the path under test.

Signature analysers can be designed to work on specific parts of the transmission only.[8] If the interface is carrying program material, the signature will vary from frame to frame. However, the ancillary data slots can still be used for signature analysis.

In component systems, the signature analyser may be set to operate on only one selected component. Some machines can be set to operate only on selected bits in the sample, making stuck bits in parallel systems very easy to find.

References

1. CCIR Recommendation 656
2. SMPTE 125M, Television – Bit Parallel Digital Interface – Component Video Signal 4:2:2
3. EBU Doc. Tech. 3246
4. SMPTE Proposed Standard – Bit parallel digital interface for 625/50 system PAL composite digital video signals
5. SMPTE 244M, Television – System M/NTSC Composite Video Signals – Bit-Parallel Digital Interface
6. SMPTE Proposed Standard – 10-bit 4:2:2 Component and 4FSc NTSC Composite Digital Signals – Serial Digital Interface
7. EGUCHI, T., Pathological check codes for serial digital interface systems. Presented at SMPTE Conf. (Los Angeles, 1991)
8. ELKIND, R. and FIBUSH, D., Proposal for error detection and handling in studio equipment. Presented at 25th SMPTE Television Conf. (Detroit, 1991)

Chapter 6

Introduction to the digital VTR

6.1 History of DVTRs

Whilst numerous experimental machines were built previously, the first production DVTR, launched in 1987, used the D-1 format which recorded colour difference data according to CCIR-601 on ¾ inch tape. Whilst it represented a tremendous achievement, the D-1 format was too early to take advantage of high-coercivity tapes and its recording density was quite low, leading to large cassettes and high running costs. The majority of broadcasters then used composite signals, and a component recorder could not easily be used in such an environment. Where component applications existed, the D-1 format could not compete economically with Betacam SP and M-II analog formats. As a result D-1 found application only in high-end post production suites.

The D-2 format came next, but this was a composite digital format, handling conventional PAL and NTSC signals in digital form, and derived from a format developed by Ampex for an automated cart. machine. The choice of composite recording was intended to allow broadcasters directly to replace analog recorders with a digital machine. D-2 retained the cassette shell of D-1 but employed higher-coercivity tape and azimuth recording (see Chapter 4) to improve recording density and playing time. Early D-2 machines had no flying erase heads, and difficulties arose with audio edits. D-2 was also hampered by the imminent arrival of the D-3 format.

D-3 was designed by NHK, and put into production by Panasonic. This had twice the recording density of D-2 and three times that of D-1. This permitted the use of ½ inch tape, making a digital camcorder a possibility. D-3 used the same sampling structure as D-2 for its composite recordings. Coming later, D-3 had learned from earlier formats and had a more powerful error-correction strategy than earlier formats, particularly in the audio recording.

By this time the economics of VLSI chips had made data reduction in VTRs viable, and the first application was the Ampex DCT format which used approximately 2:1 data reduction so that component video could be recorded on an updated version of the ¾ inch cassettes and transports designed for D-2.

When Sony were developing the Digital Betacam format, compatibility with the existing analog Betacam format was a priority. Digital Betacam uses the same cassette shells as the analog format, and certain models of the digital recorder can play existing analog tapes. Sony also adopted data reduction, but this was in order to allow the construction of a digital component VTR which offered sufficient playing time within the existing cassette dimensions.

The D-5 component format is backward compatible with D-3. The same cassettes are used and D-5 machines can play D-3 tapes. However, data reduction is not used; the tape speed is doubled in the component format in order to increase the bit rate.

In the future, recording technology will continue to advance and further formats are inevitable as manufacturers perceive an advantage over their competition. This does not mean that the user need slavishly change to every new format, as the cost of format change is high. The astute user retains his or her current format for long enough to allow a number of new formats to be introduced, then makes a quantum leap to a format which is much better than the present one, missing out those between and minimizing the changeover costs.

6.2 The rotary-head tape transport

The high bit rate of digital video could be accommodated by a conventional tape deck having many parallel tracks, but each would need its own read/write electronics and the cost would be high. However, the main problem with such an approach is that the data rate is proportional to the tape speed. The provision of stunt modes such as still frame or picture-in-shuttle are difficult or impossible. The rotary-head recorder has the advantage that the spinning heads create a high head-to-tape speed offering a high bit rate recording with a small number of heads and without high tape speed. The head-to-tape speed is dominated by the rotational speed, and the linear tape speed can vary enormously without changing the frequencies produced by the head by very much. Whilst mechanically complex, the rotary-head transport has been raised to a high degree of refinement and offers the highest recording density and thus lowest cost per bit of all digital recorders. Figure 6.1 shows that the tape is led around a rotating drum in a helix such that the entrance and exit heights are different. As a result the rotating heads cross the tape at an angle and record a series of slanting tracks. The rotating heads turn at a speed which is locked to the video field rate so that a whole number of tracks results in each input field. Time compression can be used so that the switch from one track to the next falls within a gap between data blocks. Clearly the slant tracks can only be played back properly if linear tape motion is controlled in some way. This is the job of the linear control track which carries a pulse corresponding to every slant track. The control track is played back in order to control the capstan. The breaking up of fields into several tracks is called segmentation and it is used to keep the tracks reasonably short. The segments are invisibly reassembled in memory on replay to restore the original fields.

Figure 6.2 shows the important components of a rotary-head helical-scan tape transport. There are four servo systems which must correctly interact to obtain all modes of operation: two reel servos, the drum servo and the capstan servo. The capstan and reel servos together move and tension the tape, and the drum servo moves the heads. For variable-speed operation a further servo system will be necessary to deflect the heads.

There are two approaches to capstan drive, those which use a pinch roller and those which do not. In a pinch roller drive, the tape is held against the capstan by pressure from a resilient roller which is normally pulled towards the capstan by a solenoid. The capstan only drives the tape over a narrow speed range, generally

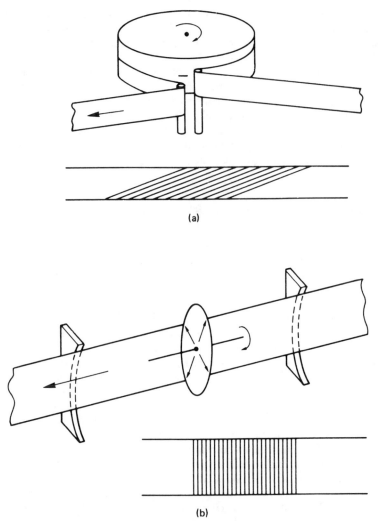

Figure 6.1 Types of rotary-head recorder. (a) Helical scan records long diagonal tracks.
(b) Transverse scan records short tracks across the tape.

the range in which broadcastable pictures are required. Outside this range, the
pinch roller retracts, the tape will be driven by reel motors alone, and the reel
motors will need to change their operating mode; one becomes a velocity servo
whilst the other remains a tension servo.

In a pinch-roller-less transport, the tape is wrapped some way around a
relatively large capstan, to give a good area of contact. The tape is always in
contact with the capstan, irrespective of operating mode, and so the reel servos
never need to change mode. A large capstan has to be used to give sufficient
contact area, and to permit high shuttle speed without excessive motor rpm. This
means that at play speed it will be turning slowly, and must be accurately

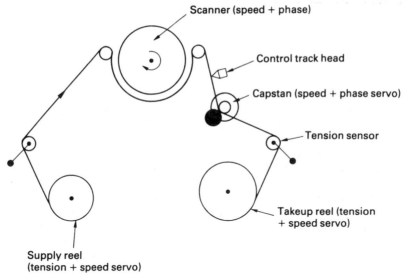

Figure 6.2 The four servos essential for proper operation of a helical-scan DVTR. Cassette-based units will also require loading and threading servos, and for variable speed a track-following servo will be necessary.

controlled and free from cogging. A multipole ironless-rotor pancake-type brush motor is often used, or a sinusoidal-drive brushless motor.

The simplest operating mode to consider is the first recording on a blank tape. In this mode, the capstan will rotate at constant speed, and drive the tape at the linear speed specified for the format. The drum must rotate at a precisely determined speed, so that the correct number of tracks per unit distance will be laid down on the tape. Since in a segmented recording each track will be a constant fraction of a television field, the drum speed must ultimately be determined by the incoming video signal to be recorded. To take the example of a PAL D-2 or D-3 recorder having two record head pairs, eight tracks or four segments will be necessary to record one field, and so the drum must make exactly two complete revolutions in one field period, requiring it to run at 100 Hz. In the case of NTSC D-2 or D-3, there are six tracks or three segments per field, and so the drum must turn at one and a half times field rate, or a little under 90 Hz. The phase of the drum rotation with respect to input video timing depends upon the time delay necessary to shuffle and interleave the video samples. This time will vary from a minimum of about one segment to more than a field depending on the format.

In order to obtain accurate tracking on replay, a phase comparison will be made between offtape control-track pulses and pulses generated by the rotation of the drum. If the phase error between these is used to modify the capstan drive, the error can be eliminated, since the capstan drives the tape which produces the control-track segment pulses. Eliminating this timing error results in the rotating heads following the tape tracks properly. Artificially delaying or advancing the reference pulses from the drum will result in a tracking adjustment.

6.3 Digital video cassettes

The D-1 and D-2 formats use the same mechanical parts and dimensions in their respective ¾ inch cassettes, even though the kind of tape and the track pattern are completely different. The Ampex DCT cassette is the same as a D-2 cassette. D-3 and D-5 use the same ½ inch cassette, and Digital Betacam uses a ½ inch cassette which is identical mechanically to the analog Betacam cassette, but uses different tape.

The main advantages of a cassette are that the medium is better protected from contamination whilst out of the transport, and that an unskilled operator or a mechanical elevator can load the tape.

The digital cassette contains two fully flanged reels side by side. The centre of each hub is fitted with a thrust pad and when the cassette is not in the drive a spring acts on this pad and presses the lower flange of each reel firmly against the body of the cassette to exclude dust. When the cassette is in the machine the relative heights of the reel turntables and the cassette supports are such that the reels seat on the turntables before the cassette comes to rest. This opens a clearance space between the reel flanges and the cassette body by compressing the springs.

The use of a cassette means that it is not as easy to provide a range of sizes as it is with open reels. Simply putting smaller reels in a cassette with the same hub spacing does not produce a significantly smaller cassette. The only solution is to specify different hub spacings for different sizes of cassette. This gives the best volumetric efficiency for storage, but it does mean that the transport must be able to reposition the reel drive motors if it is to play more than one size of cassette.

Digital Betacam offers two cassette sizes, whereas the other formats offer three. If the small, medium and large digital video cassettes are placed in a stack

		D-1	D-2	D-3	D-5	DCT	Digital Betacam
Track pitch (μm)		45	35	18	18	35	26
Tape speed (mm/s)		286·9	131·7	83·2	167·2	131·7	96·7
Play time (min)	S	14/11	32	64/50	32/25	32	40
	M	50/37	104	125/95	62/47	104	–
	L	101/75	208	245/185	123/92	208	124
Data rate (mbits/s)		216	142	142	288	113	126
Density (mbits/cm²)		4	5·8	13	13	5·8	10

Figure 6.3 The D-1/D-2, D-3/D-5, DCT and Digital Betacam tape sizes and playing times contrasted.

with their throats and tape guides in a vertical line, the centres of the hubs will be seen to fall on a pair of diagonal lines going outwards and backwards. This arrangement was chosen to allow the reel motors to travel along a linear track in machines which accept more than one size. Figure 6.3 compares the sizes and capacities of the various digital cassettes.

6.4 DVTR block diagram

Figure 6.4(a) shows a representative block diagram of a full bit rate DVTR. Following the converters will be the distribution of odd and even samples and a shuffle process for concealment purposes. An interleaved product code will be formed prior to the channel-coding stage which produces the recorded waveform. On replay the data separator decodes the channel code and the inner and outer codes perform correction as in Section 4.24. Following the de-shuffle the data channels are recombined and any necessary concealment will take place. Figure 6.4(b) shows the block diagram of a DVTR using data reduction. Data from the converters is rearranged from the normal raster scan to sets of pixel blocks upon which the data reduction unit works. A common size is eight pixels horizontally by four vertically. The blocks are then shuffled for concealment purposes. The shuffled blocks are passed through the data reduction unit. The output of this is distributed and then assembled into product codes and channel coded as for a conventional recorder. On replay data separation and error correction takes place as before, but there is now a matching data expansion unit which outputs pixel blocks. These are then de-shuffled prior to the error-concealment stage. As concealment is more difficult with pixel blocks, data from another field may be employed for concealment as well as data within the field.

The various DVTR formats largely employ the same processing stages, but there are considerable differences in the order in which these are applied. Distribution is shown in Figure 6.5(a). This is a process of sharing the input bit rate over two or more signal paths so that the bit rate recorded in each is reduced. The data are subsequently recombined on playback. Each signal path requires its own tape track and head. The parallel tracks which result form a *segment*.

Segmentation is shown in Figure 6.5(b). This is the process of sharing the data resulting from one video field over several segments. The replay system must have some means to ensure that associated segments are reassembled into the original field. This is generally a function of the control track.

Figure 6.5(c) shows a product code. Data to be recorded are protected by two error-correcting code word systems at right angles; the inner code and the outer code (see Chapter 4). When it is working within its capacity the error-correction system returns corrupt data to its original value and its operation is undetectable.

If errors are too great for the correction system, concealment will be employed. Concealment is the *estimation* of missing data values from surviving data nearby. Nearby means data on vertical, horizontal or time axes as shown in Figure 6.5(d). Concealment relies upon distribution, as all tracks of a segment are unlikely to be simultaneously lost, and upon the *shuffle* shown in Figure 6.5(e). Shuffling reorders the pixels prior to recording and is reversed on replay. The result is that uncorrectable errors due to dropouts are not concentrated, but are spread out by the de-shuffle, making concealment easier. A different approach is required where data reduction is used because the data recorded are not pixels

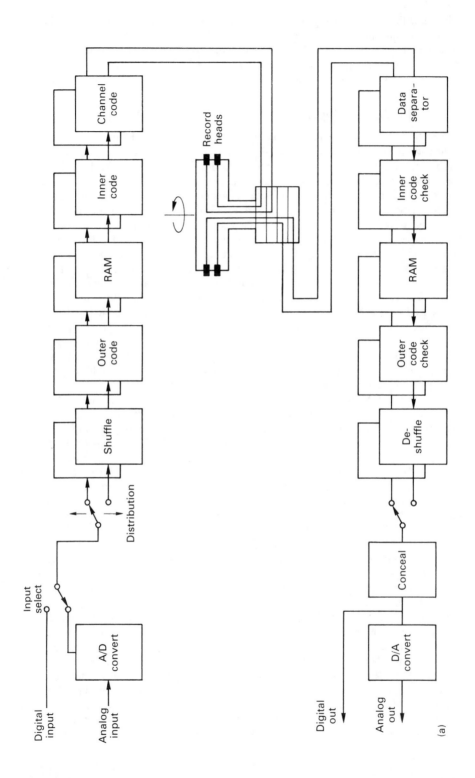

(a)

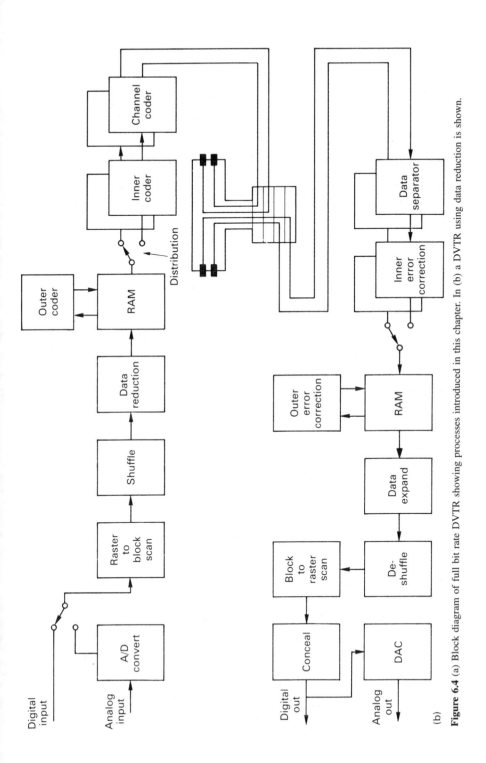

Figure 6.4 (a) Block diagram of full bit rate DVTR showing processes introduced in this chapter. In (b) a DVTR using data reduction is shown.

representing a point, but coefficients representing an area of the image. In this case it is the DCT blocks (typically eight pixels across) which must be shuffled.

There are two approaches to error correction in segmented recordings. In D-1 and D-2 the approach shown in Figure 6.6(a) is used. Here, following distribution the input field is segmented first, then each segment becomes an independent shuffled product code. This requires less RAM to implement, but it means that from an error-correction standpoint each tape track is self-contained and must deal alone with any errors encountered.

Later formats, beginning with D-3, use the approach shown in Figure 6.6(b). Here, following distribution the entire field is used to produce one large shuffled product code in each channel. The product code is then segmented for recording on tape. Although more RAM is required to assemble the large product code, the result is that outer codewords on tape spread across several tracks and

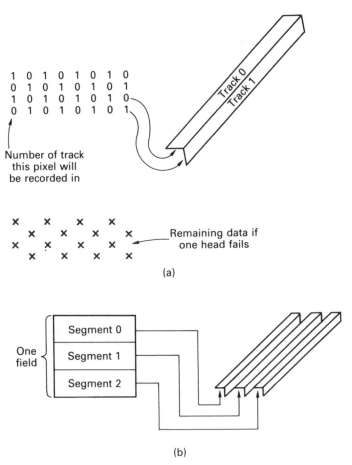

Figure 6.5 The fundamental stages of DVTR processing. In (a), distribution spreads data over more than one track to make concealment easier and to reduce the data rate per head. In (b) segmentation breaks video fields into manageable track lengths.

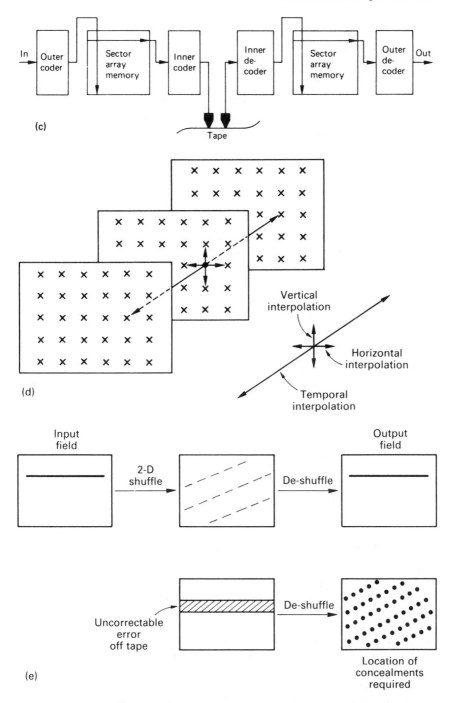

Figure 6.5 continued Product codes (c) correct mixture of random and burst errors. Correction failure requires concealment which may be in three dimensions as shown in (d). Irregular shuffle (e) makes concealments less visible.

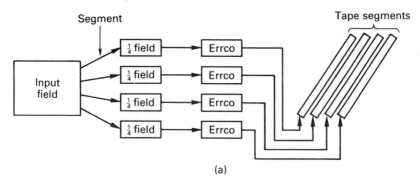

(a)

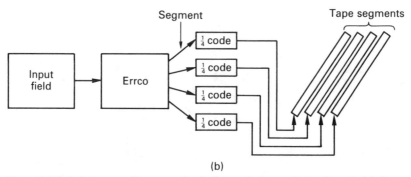

(b)

Figure 6.6 Early formats would segment data before producing product codes as in (a). Later formats perform product coding first, and then segment for recording as in (b). This gives more robust performance.

redundancy in one track can compensate for errors in another. The result is that size of a single burst error which can be fully corrected is increased. As RAM is now cheaper than when the first formats were designed, this approach is becoming more common.

6.5 Operating modes of a DVTR

A simple recorder needs to do little more than record and play back, but the sophistication of modern production processes requires a great deal more flexibility. A production DVTR will need to support most if not all of the following:

● Offtape confidence replay must be available during recording
● Timecode must be recorded and this must be playable at all tape speeds. Full remote control is required so that edit controllers can synchronize several machines via timecode.
● A high quality video signal is required over a speed range of $-1 \times$ to $+3 \times$ normal speed. Audio recovery is required in order to locate edit points from dialogue.

- A picture of some kind is required over the whole shuttle speed range (typically ±50 × normal speed).
- Assembly and insert editing must be supported, and it must be possible to edit the audio and video channels independently.
- It must be possible to change the replay speed slightly in order to shorten or lengthen programs. Full audio and video quality must be available in this mode, known as tape speed override (TSO)
- For editing purposes, there is a requirement for the DVTR to be able to play back the tape with heads fitted in advance of the record head. The playback signal can be processed in some way and re-recorded in a single pass. This is known as preread, or read–modify–write operation.

6.6 Confidence replay

It is important to be quite certain that a recording is being made, and the only way of guaranteeing that this is so is actually to play the tape as it is being recorded. Extra heads are fitted to the revolving drum in such a way that they pass along the slant tracks directly behind the record heads. The drum must carry additional rotary transformers so that signals can simultaneously enter and leave the rotating head assembly. As can be seen in Figure 6.7 the input signal will be made available at the machine output if all is well. In analog machines it was traditional to assess the quality of the recording by watching a picture monitor connected to the confidence replay output during recording. With a digital machine this is not necessary, and instead the rate at which the replay channel performs error corrections should be monitored. In some machines the error rate is made available at an output socket so that remote or centralized data reliability logging can be used.

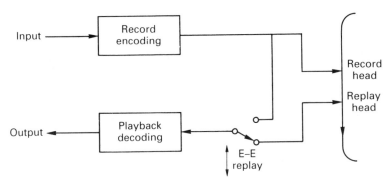

Figure 6.7 A professional DVTR uses confidence replay where the signal recorded is immediately played back. If the tape is not running, the heads are bypassed in E–E mode so that all of the circuitry can be checked.

It will be seen from Figure 6.7 that when the machine is not running, a connection is made which bypasses the record and playback heads. The output signal in this mode has passed through every process in the machine except the actual tape/head system. This is known as E–E (electronics to electronics) mode, and is a good indication that the circuitry is functioning.

6.7 Colour framing

As was seen in Chapter 5, composite video has a subcarrier added to the luminance in order to carry the colour-difference signals. The frequency of this subcarrier is critical if it is to be invisible on monochrome TV sets, and as a result it does not have a whole number of cycles within a frame, but only returns to its starting phase once every two frames in NTSC and every four frames in PAL. These are known as colour-framing sequences. When playing back a composite recording, the offtape colour-frame sequence must be synchronized with the reference colour-frame sequence, otherwise composite replay signals cannot be mixed with signals from elsewhere in the facility. When editing composite recordings, the subcarrier phase must not be disturbed at the edit point and this puts constraints on the edit control process. In both cases the solution is to record the start of a colour-frame sequence in the control track of the tape. There is also a standardized algorithm linking the timecode with the colour-framing sequences.

6.8 Timecode

Timecode is simply a label attached to each frame on the tape which contains the time at which it was recorded measured in hours, minutes, seconds and frames. There are two ways in which timecode data can be recorded. The first is to use a dedicated linear track, usually alongside the control track, in which there is one timecode entry for every tape frame. Such a linear track can easily be played back over a wide speed range by a stationary head. Timecode of this kind is known as LTC (linear timecode).

LTC clearly cannot be replayed when the tape is stopped or moving very slowly. In DVTRs with track-following heads, particularly those which support preread, head deflection may result in a frame being played which is not the one corresponding to the timecode from the stationary head. The player software needs to modify the LTC value as a function of the head deflection.

An alternative timecode is where the information is recorded in the video field itself, so that the above mismatch cannot occur. This is known as vertical interval time code (VITC) because it is recorded in a line which is in the vertical blanking period. VITC has the advantage that it can be recovered with the tape stopped, but it cannot be recovered in shuttle because the rotary heads do not play entire tracks in this mode. DVTRs do not record the whole of the vertical blanking period, and if VITC is to be used, it must be inserted in a line which is within the recorded data area of the format concerned.

6.9 Picture-in-shuttle

A rotary-head recorder cannot follow the tape tracks properly when the tape is shuttled. Instead the heads cross the tracks at an angle and intermittently pick up short data blocks. Each of these blocks is an inner error-correcting codeword and this can be checked to see if the block was properly recovered. If this is the case, the data can be used to update a frame store which displays the shuttle picture. Clearly the shuttle picture is a mosaic of parts of many fields. In addition to helping the concealment of erors, the shuffle process is beneficial to obtaining picture-in-shuttle. Owing to shuffle, a block recovered from the tape contains data from many places in the picture, and this gives a better result than if many pixels were available from one place in the picture. The twinkling effect seen in shuttle is due to the updating of individual pixels following de-shuffle.

When data reduction is used, the picture is processed in blocks, and these will be visible as mosaicing in the shuttle picture as the frame store is updated by the blocks.

In composite recorders, the subcarrier sequence is only preserved when playing at normal speed. In all other cases, extra *colour processing* is required to convert the disjointed replay signal into a continuous subcarrier once more.

6.10 Digital Betacam

Digital Betacam (DB) is a component format which accepts 8 or 10 bit 4:2:2 data with 720 luminance samples per active line and four channels of 48 kHz digital audio having up to 20 bit wordlength. Video data reduction based on discrete cosine transform is employed, with a compression factor of almost two to one (assuming 8 bit input). The audio data are uncompressed. The cassette shell of the ½ inch analog Betacam format is retained, but contains 14 micrometre metal particle tape. The digital cassette contains an identification hole which allows the transport to identify the tape type. Unlike the other digital formats, only two cassette sizes are available. The large cassette offers 124 minutes of playing time; the small cassette plays for 40 minutes.

Owing to the trade-off between SNR and bandwidth which is a characteristic of digital recording, the tracks must be longer than in the analog Betacam format, but narrower. The drum diameter of the DB transport is 81.4 mm which is designed to produce tracks of the required length for digital recording. The helix

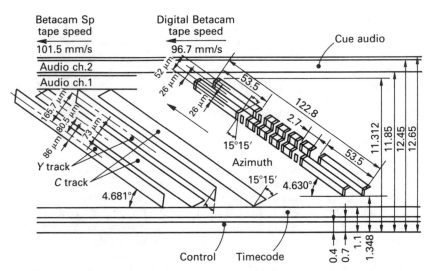

Figure 6.8 The track pattern of Digital Betacam. Control and timecode tracks are identical in location to the analog format, as is the single analog audio cue track. Note the use of a small guard band between segments.

angle of the digital drum is designed such that when an analog Betacam tape is driven past at the correct speed, the track angle is correct. Certain DB machines are fitted with analog heads which can trace the tracks of an analog tape. As the drum size is different, the analog playback signal is time compressed by about 9%, but this is easily dealt with in the timebase-correction process.[1] The fixed heads are compatible with the analog Betacam positioning. The reverse compatibility is for playback only; the digital machine cannot record on analog cassettes.

Figure 6.8 shows the track pattern for 625/50 Betacam. The four digital audio channels are recorded in separate sectors of the slant tracks, and so one of the linear audio channels of the analog format is dispensed with, leaving one linear audio track for cueing.

Azimuth recording is employed, with two tracks being recorded simultaneously by adjacent heads. Electronic delays are used to align the position of the edit gaps in the two tracks of a segment, allowing double-width flying erase heads to be used. Three segments are needed to record one field, requiring one and a half drum revolutions. Thus the drum speed is three times that of the analog format. However, the track pitch is less than one-third that of the analog format, so the linear speed of the digital tape is actually slower. The track width is 24 micrometres, with a 4 micrometre guard band between segments making the effective track pitch 26 micrometres, compared with 18 for D-3/D-5 and 39 for D-2/DCT.

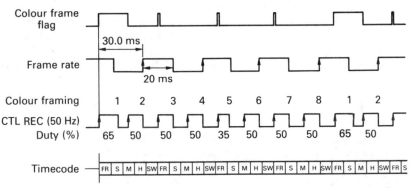

Figure 6.9 The control track of Digital Betacam uses duty cycle modulation for colour framing purposes.

There is a linear timecode track whose structure is identical to the analog Betacam timecode, and a control track shown in Figure 6.9 having a fundamental frequency of 50 Hz. Ordinarily the duty cycle is 50%, but this changes to 65/35 in field 1 and 35/65 in field 5. The rising edge of the CTL signal coincides with the first segment of a field, and the duty cycle variations allow four- or eight-field colour framing if decoded composite sources are used. As the drum speed is 75 Hz, CTL and drum phase coincide every three revolutions.

Figure 6.10 shows the track layout in more detail. Unlike other digital formats, DB incorporates tracking pilot tones recorded between the audio and video sectors. The first tone has a frequency of approximately 4 MHz and appears once

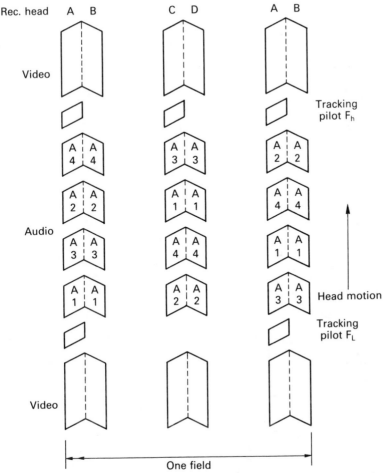

Rec. head

Figure 6.10 The sector structure of Digital Betacam. Note the tracking tones between the audio and video sectors which are played back for alignment purposes during insert edits.

per drum revolution. The second is recorded at approximately 400 kHz and appears twice per drum revolution. The pilot tones are recorded when a recording is first made on a blank tape, and will be re-recorded following an assemble edit; however, during an insert edit the tracking pilots are not re-recorded, but used as a guide to the insertion of the new tracks.

The amplitude of the pilot signal is a function of the head tracking. The replay heads are somewhat wider than the tracks, and so a considerable tracking error will have to be present before a loss of amplitude is noted. This is partly offset by the use of a very-long-wavelength pilot tone in which fringing fields increase the effective track width. The low-frequency tones are used for automatic playback tracking.

With the tape moving at normal speed, the capstan phase is changed by steps in one direction and then the other as the pilot tone amplitude is monitored. The phase which results in the largest amplitude will be retained. During the edit

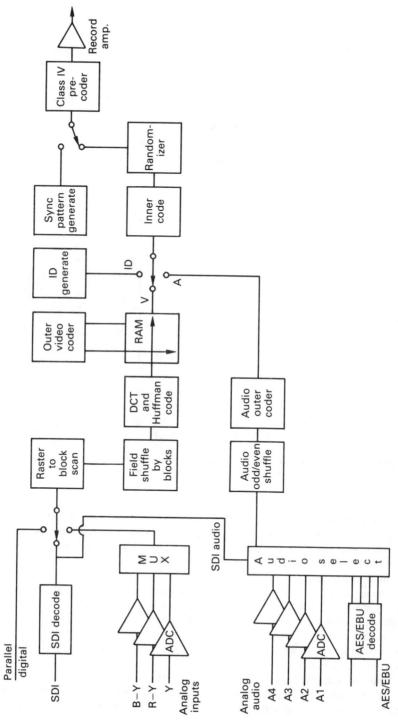

Figure 6.11 Block diagram of Digital Betacam record channel. Note that the use of data reduction makes this rather different to the block layout of full-bit formats.

preroll the record heads play back the high-frequency pilot tone and capstan phase is set for largest amplitude. The record heads are the same width as the tracks and a short-wavelength pilot tone is used such that any mistracking will cause immediate amplitude loss. This is an edit optimize process and results in new tracks being inserted in the same location as the originals. As the pilot tones are played back in insert editing, there will be no tolerance build-up in the case of multiple inserts.

The data reduction of DB^2 works on an intrafield basis to allow complete editing freedom and uses processes described in Chapter 3.

Figure 6.11 shows a block diagram of the record section of DB. Analog component inputs are sampled at 13.5 and 6.75 MHz. Alternatively the input may be SDI at 270 Mbits/s which is deserialized and demultiplexed to separate components. The raster scan input is first converted to blocks which are 8 pixels wide by 4 pixels high in the luminance channel and 4 pixels by 4 in the two colour difference channels. When two fields are combined on the screen, the result is effectively an interlaced 8 × 8 luminance block with colour difference pixels having twice the horizontal luminance pixel spacing. The pixel blocks are then subject to a field shuffle. A shuffle based on individual pixels is impossible because it would raise the high-frequency content of the image and destroy the power of the data reduction process. Instead the block shuffle helps the data reduction by making the average entropy of the image more constant. This happens because the shuffle exchanges blocks from flat areas of the image with blocks from highly detailed areas. The shuffle algorithm also has to consider the requirements of picture-in-shuttle. The blocking and shuffle take place when the read addresses of the input memory are permuted with respect to the write addresses.

Following the input shuffle the blocks are associated into sets of ten in each component and are then subject to the discrete cosine transform. The resulting coefficients are then subject to an iterative requantizing process followed by variable-length coding. The iteration adjusts the size of the quantizing step until the overall length of the ten coefficient sets is equal to the constant capacity of an entropy block which is 364 bytes. Within that entropy block the amount of data representing each individual DCT block may vary considerably, but the overall block size stays the same.

The DCT process results in coefficients whose wordlength exceeds the input wordlength. As a result it does not matter if the input wordlength is 8 bits or 10 bits; the requantizer simply adapts to make the output data rate constant. Thus the compression is greater with 10 bit input, corresponding to about 2.4 to 1.

The next step is the generation of product codes as shown in Figure 6.12(a). Each entropy block is divided into two halves of 162 bytes each and loaded into the rows of the outer code RAM which holds 114 such rows, corresponding to one-twelfth of a field. When the RAM is full, it is read in columns by address mapping and 12 bytes of outer Reed–Solomon redundancy are added to every column, increasing the number of rows to 126.

The outer code RAM is read out in rows once more, but this time all 126 rows are read in turn. To the contents of each row is added a 2 byte ID code and then the data plus ID bytes are turned into an inner code by the addition of 14 bytes of Reed–Solomon redundancy.

Inner codewords pass through the randomizer and are then converted to serial form for Class IV partial response precoding. With the addition of a sync pattern

(a) Video ... 12 ECC blocks/field (2 ECC blocks/track)

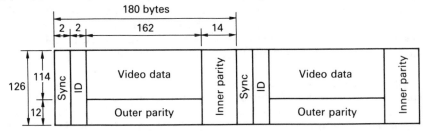

(b) Audio ... 2 ECC blocks/ (CH. × field)

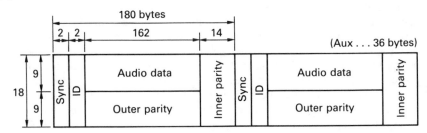

(c) Sync block

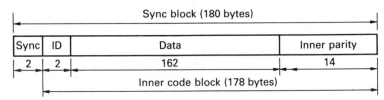

Figure 6.12 Video product codes of Digital Betacam are shown in (a); 12 of these are needed to record one field. Audio product codes are shown in (b); two of these record samples corresponding to one field period. Sync blocks are common to audio and video as shown in (c). The ID code discriminates between video and audio channels.

of two bytes, each inner codeword becomes a sync block as shown in Figure 6.12(c). Each video block contains 126 sync blocks, preceded by a preamble and followed by a postamble. One field of video data requires twelve such blocks. Pairs of blocks are recorded simultaneously by the parallel heads of a segment. Two video blocks are recorded at the beginning of the track, and two more are recorded after the audio and tracking tones.

The audio data for each channel are separated into odd and even samples for concealment purposes and assembled in RAM into two blocks corresponding to one field period. Two 20 bit samples are stored in 5 bytes. Figure 6.12(b) shows that each block consists of 1458 bytes including auxiliary data from the AES/

EBU interface, arranged as a block of 162 × 9 bytes. 100% outer code redundancy is obtained by adding 9 bytes of Reed–Solomon check bytes to each column of the blocks.

The inner codes for the audio blocks are produced by the same circuitry as the video inner codes on a time-shared basis. The resulting sync blocks are identical in size and differ only in the provision of different ID codes. The randomizer and precoder are also shared. It will be seen from Figure 6.10 that there are three segments in a field and that the position of an audio sector corresponding to a particular audio channel is different in each segment. This means that damage due to a linear tape scratch is distributed over three audio channels instead of being concentrated in one.

Each audio product block results in 18 sync blocks. These are accommodated in audio sectors of six sync blocks each in three segments. The audio sectors are preceded by preambles and followed by postambles. Between these are edit gaps which allow each audio channel to be independently edited.

By spreading the outer codes over three different audio sectors the correction power is much improved because data from two sectors can be used to correct errors in the third.

Figure 6.13 shows the replay channel of DB. The RF signal picked up by the replay head passes first to the Class IV partial response playback circuit in which it becomes a three-level signal as was shown in Chapter 4. The three-level signal is passed to an ADC which converts it into a digitally represented form so that the Viterbi detection can be carried out in logic circuitry. The sync detector identifies the synchronizing pattern at the beginning of each sync block and resets the block bit count. This allows the entire inner codeword of the sync block to deserialized into bytes and passed to the inner error checker. Random errors will be corrected here, whereas burst errors will result in the block being flagged as in error.

Sync blocks are written into the de-interleave RAM with error flags where appropriate. At the end of each video sector the product code RAM will be filled, and outer code correction can be performed by reading the RAM at right angles and using the error flags to initiate correction by erasure. Following outer code correction the RAM will contain corrected data or uncorrectable error flags which will later be used to initiate concealment.

The sync blocks can now be read from memory and assembled in pairs into entropy blocks. The entropy block is of fixed size, but contains coefficient blocks of variable length. The next step is to identify the individual coefficients and separate the luminance and colour difference coefficients by decoding the Huffman-coded sequence. Following the assembly of coefficient sets the inverse DCT will result in pixel blocks in three components once more. The pixel blocks are de-shuffled by mapping the write address of a field memory. When all of the tracks of a field have been decoded, the memory will contain a de-shuffled field containing either correct sample data or correction flags. By reading the memory without address mapping the de-shuffled data are then passed through the concealment circuit where flagged data are concealed by data from nearby in the same field or from a previous field. The memory readout process is buffered from the offtape timing by the RAM and as a result the timebase-correction stage is inherent in the replay process. Following concealment the data can be output as conventional raster scan video either formatted to parallel or serial digital standards or converted to analog components.

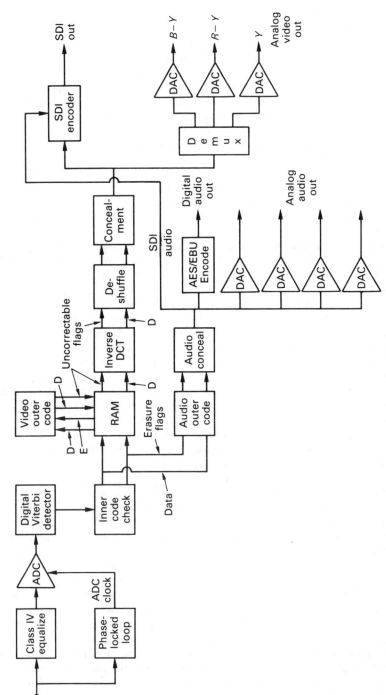

Figure 6.13 The replay channel of Digital Betacam. This differs from a full-bit system primarily in the requirement to deserialize variable-length coefficient blocks prior to the inverse DCT.

6.11 Digital audio in VTRs

The audio performance of video recorders has traditionally lagged behind that of audio-only recorders. In video recorders, the use of rotary heads to obtain sufficient bandwidth results in a wide tape whose longitudinal audio tracks travel relatively slowly by professional audio standards. In all rotary-head recorders, the intermittent head contact causes shock-wave patterns to propagate down the tape, making low-flutter figures difficult to achieve. This is compounded by the action of the capstan servo which has to change tape speed to maintain control-track phase if the video heads are to track properly.

The requirements of lip-sync dictate that the same head must be used for both recording and playback, when the optimum head design for these two functions is different. When dubbing from one track to the next, one head gap will be recording the signal played back by the adjacent magnetic circuit in the head, and mutual inductance can cause an oscillatory loop if extensive antiphase crosstalk cancelling is not employed. Placing the tracks on opposite sides of the tape would help this problem, but phase errors between the channels can then be introduced by tape weave. This can mean the difference between a two-channel recorder and a stereo recorder. Crosstalk between the timecode and audio tracks can also restrict performance.

The adoption of digital techniques essentially removes these problems for the audio in a DVTR. Once the audio is in numerical form, wow, flutter and channel-phase errors can be eliminated by timebase correction; crosstalk ceases to occur and, provided a suitable error-correction strategy is employed, the only degradation of the signal will be due to quantizing. The most significant advantages of digital recording are that there is essentially no restriction on the number of generations of re-recording which can be used and that proper crossfades can be made in the audio at edit points, following a rehearsal if necessary.

Digital audio recording with video is rather more difficult than in an audio-only environment.

The audio samples are carried by the same channel as the video samples. The audio could have used separate stationary heads, but this would have increased tape consumption and machine complexity. In order to permit independent audio and video editing, the tape tracks are given a block structure. Editing will require the heads momentarily to go into record as the appropriate audio block is reached. Accurate synchronization is necessary if the other parts of the recording are to remain uncorrupted. The sync block structure of the video sector continues in the audio sectors because the same read/write circuitry is used for audio and video data. Clearly the ID code structure must also continue through the audio. In order to prevent audio samples from arriving in the frame store in shuttle, the audio addresses are different from the video addresses.

Despite the additional complexity of sharing the medium with video, the professional DVTR must support such functions as track bouncing, synchronous recording and split audio edits with variable crossfade times.

The audio samples in a DVTR are binary numbers just like the video samples, and although there is an obvious difference in sampling rate and wordlength, the use of time compression means that this only affects the relative areas of tape devoted to the audio and video samples. The most important difference between audio and video samples is the tolerance to errors. The acuity of the ear means

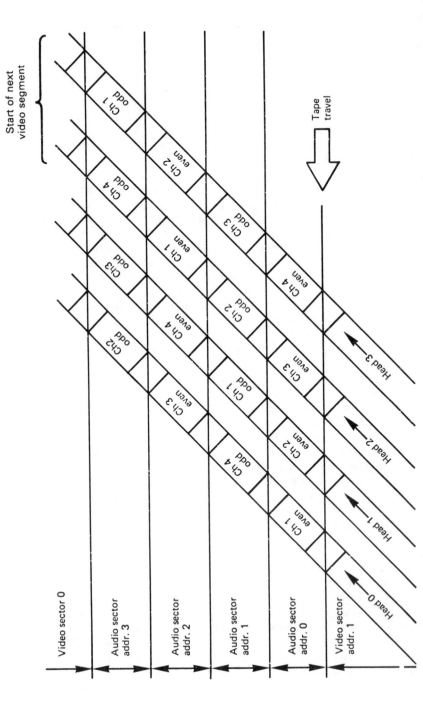

Figure 6.14 The structure of the audio blocks in D-1 format showing the double recording, the odd/even interleave, the sector addresses, and the distribution of audio channels over all heads. The audio samples recorded in this area represent a 6.666 ms time slot in the audio recording.

that uncorrected audio samples must not occur more than once every few hours. There is little redundancy in sound when compared with video, and concealment of errors is not desirable on a routine basis. In video, the samples are highly redundant, and concealment can be effected using samples from previous or subsequent lines or, with care, from the previous frame.

Whilst subjective considerations require greater data reliability in the audio samples, audio data form a small fraction of the overall data and it is difficult to protect them with an extensive interleave whilst still permitting independent editing. For these reasons major differences can be expected between the ways that audio and video samples are handled in a digital video recorder. One such difference is that the error-correction strategy for audio samples uses a greater amount of redundancy. Whilst this would cause a serious playing time penalty in an audio recorder, even doubling the audio data rate in a video recorder only raises the overall data rate by a few per cent. The arrangement of the audio blocks is also designed to maximize data integrity in the presence of tape defects and head clogs. The audio blocks are at the ends of the head sweeps in D-2 and D-3, but are placed in the middle of the segment in D-1, D-5, Digital Betacam and DCT.

The audio sample interleave varies in complexity between the various formats. It will be seen from Figure 6.14 that the physical location of a given audio channel rotates from segment to segment. In this way a tape scratch will cause slight damage to all channels instead of serious damage to one. Data are also distributed between the heads, and so if one (or sometimes two) of the heads clogs the audio is still fully recovered.

6.12 Writing audio sectors

In each sector, the track commences with a preamble to synchronize the phase-locked loop in the data separator on replay. Each of the sync blocks begins, as the name suggests, with a synchronizing pattern which allows the read sequencer to deserialize the block correctly. At the end of a sector, it is not possible simply to turn off the write current after the last bit, as the turnoff transient would cause data corruption. It is necessary to provide a postamble such that current can be turned off away from the data. It should now be evident that any editing has to take place a sector at a time. Any attempt to rewrite one sync block would result in damage to the previous block owing to the physical inaccuracy of replacement, damage to the next block due to the turnoff transient, and inability to synchronize to the replaced block because of the random phase jump at the point where it began. The sector in a DVTR is analogous to the cluster in a disk drive. Owing to the difficulty of writing in exactly the same place as a previous recording, it is necessary to leave tolerance gaps between sectors where the write current can turn on and off to edit individual write blocks. For convenience, the tolerance gaps are made the same length as a whole number of sync blocks. Figure 6.15 shows that in D-1 the edit gap is two sync blocks long, as it is in D-3, whereas in D-2 it is only one sync block long. The first half of the tolerance gap is the postamble of the previous block, and the second half of the tolerance gap acts as the preamble for the next block. The tolerance gap following editing will contain, somewhere in the centre, an arbitrary jump in bit phase, and a certain amount of corruption due to turnoff transients. Provided that the postamble and preamble remain intact, this is of no consequence.

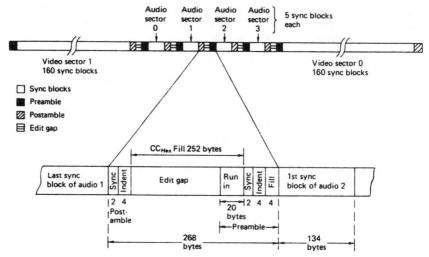

Figure 6.15 The position of preambles and postambles with respect to each sector is shown along with the gaps necessary to allow individual sectors to be written without corrupting others. When the whole track is recorded, 252 bytes of CC hex fill are recorded after the postamble before the next sync pattern. If a subsequent recording commences in this gap, it must do so at least 20 bytes before the end in order to write a new run-in pattern for the new recording.

6.13 Synchronization between audio sampling rate and video field rate

Clearly the number of audio sync blocks in a given time is determined by the number of video fields in that time. It is only possible to have a fixed tape structure if the audio sampling rate is locked to video. With 625/50 machines, the sampling rate of 48 kHz results in exactly 960 audio samples in every field.

For use on 525/60, it must be recalled that the 60 Hz is actually 59.94 Hz. As this is slightly slow, it will be found that in sixty fields, exactly 48 048 audio samples will be necessary. Unfortunately 60 will not divide into 48 048 without a remainder. The largest number which will divide 60 and 48 048 is 12; thus in 60/12 = 5 fields there will be 48 048/12 = 4004 samples. Over a five-field sequence the fields contain 801, 801, 801, 801 and 800 samples respectively, adding up to 4004 samples.

6.14 AES/EBU compatibility

In order to comply with the AES/EBU digital audio interconnect, wordlengths between 16 and 20 bits can be supported, but it is necessary to record a code in the sync block to specify the wordlength in use. Pre-emphasis may have been used prior to conversion, and this status is also to be conveyed, along with the four channel-use bits. The AES/EBU digital interconnect uses a block-sync pattern which repeats after 192 sample periods corresponding to 4 ms at 48 kHz. Since the block size is different to that of the DVTR interleave block, there can be any phase relationship between interleave block boundaries and the AES/EBU block-sync pattern. In order to recreate the same phase relationship between block sync and sample data on replay, it is necessary to record the position of block sync within the interleave block. It is the function of the interface control word in the audio data to convey these parameters. There is no guarantee that the 192 sample block-sync sequence will remain intact after audio editing; most likely there will be an arbitrary jump in block-sync phase. Strictly speaking a DVTR playing back an edited tape would have to ignore the block-sync positions on the tape, and create new block sync at the standard 192 sample spacing. Unfortunately the DVTR formats are not totally transparent to the whole of the AES/EBU data stream, as certain information is not recorded.

References

1. HUCKFIELD, D., SATO, N. and SATO, I., Digital Betacam – The application of state of the art technology to the development of an affordable component DVTR. *Record of 18th ITS*, 180–199 (Montreux, 1993)
2. CREED, D. and KAMINAGA, K. Digital compression strategies for video tape recorders in studio applications. *Record of 18th ITS*, 291–301 (Montreux, 1993)

Chapter 7

Non-linear video editing

Non-linear editing takes advantage of the freedom to store digitized image data in any suitable medium and the signal processing techniques developed in computation. The images may have originated on film or video. Recently images which have been synthesized by computer have been added. Although aesthetically film and video have traditionally had little in common, from a purely technological standpoint many of the necessary processes are similar.

7.1 Introduction

In all types of editing the goal is the appropriate sequence of material at the appropriate time. In an ideal world the difficulty and cost involved in creating the perfect edited work are discounted. In practice there is economic pressure to speed up the editing process and to use cheaper media. Editors will not accept new technologies if they form an obstacle to the creative process, but if a new approach to editing takes nothing away, it will be considered. If something is added, such as freedom or flexibility, so much the better.

When there was only film or videotape editing, it did not need a qualifying name. Now that images are stored as data, alternative storage media have become available which allow editors to reach the same goal but using different techniques. Whilst digital VTR formats copy their analog predecessors and support field-accurate editing on the tape itself, in all other digital editing samples from various sources are brought from the storage media to various pages of RAM. The edit is viewed by selectively processing two (or more) sample streams retrieved from RAM. Thus the nature of the storage medium does not affect the form of the edit in any way except the amount of time needed to execute it.

Tapes only allow serial access to data, whereas disks and RAM allow random access and so can be much faster. Editing using random access storage devices is very powerful as the shuttling of tape reels is avoided. The technique is sometimes called non-linear editing. This is not a very helpful name, as in these systems the editing itself is performed in RAM in the same way as before. In fact it is only the time axis of the storage medium which is non-linear.

7.2 The structure of a workstation

Figure 7.1 shows the general arrangement of a hard-disk-based workstation. The VDU in such devices has a screen which is a montage of many different signals,

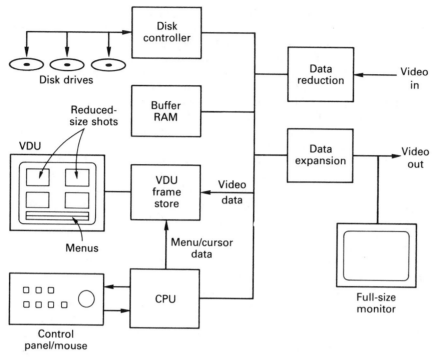

Figure 7.1 A hard-disk-based workstation. Note the screen which can display numerous clips at the same time.

each of which appear in windows. In addition to the video windows there will be a number of alphanumeric and graphic display areas required by the control system. There will also be a cursor which can be positioned by a trackball or mouse. The screen is refreshed by a frame store which is read at the screen refresh rate. The frame store can be simultaneously written by various processes to produce a windowed image. In addition to the VDU, there may be a second screen which reproduces full-size images for preview purposes.

A master timing generator provides reference signals to synchronize the internal processes. This also produces an external reference to which source devices such as VTRs can lock. The timing generator may free-run in a standalone system, or genlock to station reference to allow playout to air.

Digital inputs and outputs are provided, along with optional converters to allow working in an analog environment. In many workstations, data reduction is employed, and the appropriate coding and decoding logic will be required adjacent to the inputs and outputs. With mild compresssion, the video output of the machine may be used directly for some purposes. This is known as *online* editing. Alternatively a high compression factor may be used, and the editor is then used only to create an edit decision list (EDL). This is known as *offline* editing. The EDL is then used to control automatic editing of the full-bandwidth source material, probably on tape.

Disk-based workstations fall into two categories depending on the relative emphasis of the vertical or horizontal aspects of the process. High end post production emphasizes the vertical aspect of the editing as a large number of layers may be used to create the output image. The length of such productions is generally quite short and so disk capacity is not an issue and data reduction will not be employed. In contrast a general-purpose editor used for television programme or film production will emphasize the horizontal aspect of the task. Extended recording ability will be needed, and data reduction is more likely.

The machine will be based around a high data rate bus, connecting the I/O, RAM, disk subsystem and the processor. If magnetic disks are used, these will be Winchester types, because they offer the largest capacity. Exchangeable magneto-optic disks may also be supported.

Before any editing can be performed, it is necessary to have source material on line. If the source material exists on MO disks with the appropriate file structure, these may be used directly. Otherwise it will be necessary to input the material in real time and record it on magnetic disks via the data-reduction system. In addition to recording the data-reduced source video, reduced-size versions of each field may also be recorded which are suitable for the screen windows.

Inputting the image data from film rushes requires telecine-to-disk transfer. Inputting from videotape requires dubbing. Both are time-consuming processes. Time can be saved by involving the disk system at an early stage. In video systems, the disk system can record camera video and timecode alongside the VTRs. Editing can then begin as soon as shooting finishes. In film work, it is possible to use video-assisted cameras where a video camera runs from the film camera viewfinder. During filming, the video is recorded on disk and both record the same timecode. Once more, editing can begin as soon as shooting is finished.

7.3 Locating the edit point

Digital editors must simulate the 'rock and roll' process of edit-point location in VTRs or flatbeds where the tape or film is moved to and fro by the action of a shuttle knob, jog wheel or joystick. Whilst DVTRs with track-following systems can work in this way, disks cannot. Disk drives transfer data intermittently and not necessarily in real time. The solution is to transfer the recording in the area of the edit point to RAM in the editor. RAM access can take place at any speed or direction and the precise edit point can then be conveniently found by monitoring signals from the RAM. In a window-based display, a source recording is attributed to a particular window, and will be reproduced within that window, with timecode displayed adjacently.

Figure 7.2 shows how the area of the edit point is transferred to the memory. The source device is commanded to play, and the operator watches the replay in the selected window. The same samples are continuously written into a memory within the editor. This memory is addressed by a counter which repeatedly overflows to give the memory a ring-like structure rather like that of a timebase corrector, but somewhat larger. When the operator sees the rough area in which the edit is required, he or she will press a button. This action stops the memory writing, not immediately, but one-half of the memory contents later. The effect is then that the memory contains an equal number of samples before and after the rough edit point.

Once the recording is in the memory, it can be accessed at leisure, and the constraints of the source device play no further part in the edit-point location.

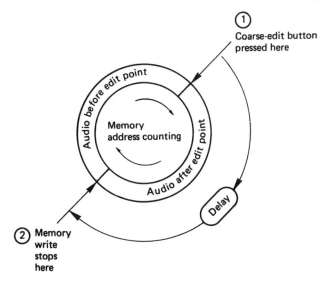

Figure 7.2 The use of a ring memory which overwrites allows storage of samples before and after the coarse edit point.

There are a number of ways in which the the memory can be read. If the field address in memory is supplied by a counter which is clocked at the appropriate rate, the edit area can be replayed at normal speed, or at some fraction of normal speed repeatedly. In order to simulate the analog method of finding an edit point, the operator is provided with a *scrub wheel* or rotor, and the memory field address will change at a rate proportional to the speed with which the rotor is turned, and in the same direction. Thus the recording can be seen forward or backward at any speed, and the effect is exactly that of turning the wheel on a flatbed or VTR.

If the position of the jog address pointer through the memory is compared with the addresses of the ends of the memory, it will be possible to anticipate that the pointer is about to reach the end of the memory. A disk transfer can be performed to fetch new data further up the time axis, so that it is possible to jog an indefinite distance along the source recording in a manner which is transparent to the user.

Samples which will be used to make the master recording need never pass through these processes; they are solely to assist in the location of the edit points.

The act of pressing the coarse edit-point button stores the timecode of the source at that point, which is frame accurate. As the rotor is turned, the memory address is monitored, and used to update the timecode.

Before the edit can be performed, two edit points must be determined, the out-point at the end of the previously recorded signal, and the in-point at the beginning of the new signal. The second edit point can be determined by moving the cursor to a different screen window in which video from a different source is displayed. The jog wheel will now roll this material to locate the second edit point while the first source video remains frozen in the deselected window. The editor's microprocessor stores these in an edit decision list (EDL) in order to control the automatic assemble process.

It is also possible to locate a rough edit point by typing in a previously noted timecode, and the image in the window will automatically jump to that time. In some systems, in addition to recording video and audio, there may also be text files locked to timecode which contain the dialog. Using these systems one can allocate a textual dialog display to a further window and scroll down the dialog or search for a key phrase as in a word processor. Unlike a word processor, the timecode pointer from the text access is used to jog the video window. As a result an edit point can be located in the video if the actor's lines at the desired point are known.

7.4 Editing with disk drives

Using one or other of the above methods, an edit list can be made which contains an in-point, an out-point and a filename for each of the segments of video which need to be assembled to make the final work, along with a timecode-referenced transition command and period for the vision mixer. This edit list will also be stored on the disk. When a preview of the edited work is required, the edit list is used to determine what files will be necessary and when, and this information drives the disk controller.

Figure 7.3 shows the events during an edit between two files. The edit list causes the relevant blocks from the first file to be transferred from disk to

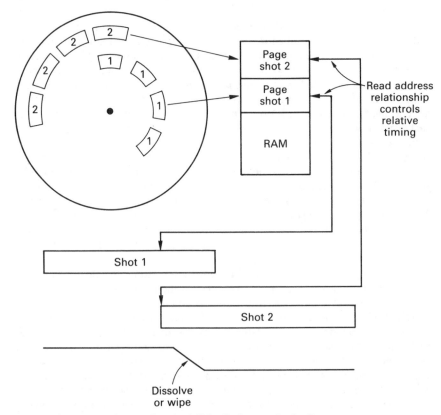

Figure 7.3 Sequence of events for a hard-disk edit. See text for details.

memory, and these will be read by the signal processor to produce the preview output. As the edit point approaches, the disk controller will also place blocks from the incoming file into the memory. In different areas of the memory there will be simultaneously the end of the outgoing recording and the beginning of the incoming recording. Before the edit point, only pixels from the outgoing recording are accessed, but as the transition begins, pixels from the incoming recording are also accessed, and for a time both data streams will be input to the vision mixer according to the transition period required. The output of the signal processor becomes the edited preview material, which can be checked for the required subjective effect. If necessary the in- or out-points can be trimmed, or the crossfade period changed, simply by modifying the edit-list file. The preview can be repeated as often as needed, until the desired effect is obtained. At this stage the edited work does not exist as a file, but is re-created each time by a further execution of the EDL. Thus a lengthy editing session need not fill up the disks.

It is important to realize that at no time during the edit process were the original files modified in any way. The editing was done solely by reading the files. The power of this approach is that if an edit list is created wrongly, the original recording is not damaged, and the problem can be put right simply by correcting the edit list. The advantage of a disk-based system for such work is that location of edit points, previews and reviews are all performed almost instantaneously, because of the random access of the disk. This can reduce the time taken to edit a program to a fraction of that needed with a tape machine.

During an edit, the disk controller has to provide data from two different files simultaneously, and so it has to work much harder than for a simple playback. If there are many close-spaced edits, the controller and drives may be hard pressed to keep ahead of real time, especially if there are long transitions, because during a transition a vertical edit is taking place between two video signals and the source data rate is twice as great as during replay. A large buffer memory helps this situation because the drive can fill the memory with files before the edit actually begins, and thus the instantaneous sample rate can be met by allowing the memory to empty during disk-intensive periods.

Some drives rotate the sector addressing from one cylinder to the next so that the drive does not lose a revolution when it moves to the next cylinder. Disk-editor performance is usually specified in terms of peak editing activity which can be achieved, but with a recovery period between edits. If an unusually severe editing task is necessary where the drive just cannot access files fast enough, it will be necessary to rearrange the files on the disk surface so that files which will be needed at the same time are on nearby cylinders. An alternative is to spread the material between two or more drives so that overlapped seeks are possible.

Once the editing is finished, it will generally be necessary to transfer the edited material to form a contiguous recording so that the source files can make way for new work. In offline editing, the source files already exist on tape or film and all that is needed is the EDL; the disk files can simply be erased. In online editing the disks hold original recordings and will need to be backed up to tape if they will be required again. In large broadcast systems, the edited work can be broadcast directly from the disk file server. In smaller systems it will be necessary to output to some removable medium, since the Winchester drives in the editor have fixed media.

7.5 Disk drives

Disk drives came into being as random access file-storage devices for digital computers. The explosion in personal computers has fuelled demand for low-cost high-density disk drives and the rapid access offered is increasingly finding applications in digital video. After lengthy development, optical disks are also emerging in digital video applications.

Figure 7.4 shows that, in a disk drive, the data are recorded on a circular track. In hard-disk drives, the disk rotates at several thousand rpm so that the head-to-disk speed is of the order of 100 miles per hour. At this speed no contact can be tolerated, and the head flies on a boundary layer of air turning with the disk at a height measured in microinches. The longest time it is necessary to wait to access a given data block is a few milliseconds. To increase the storage capacity of the drive without a proportional increase in cost, many concentric tracks are recorded on the disk surface, and the head is mounted on a positioner which can rapidly bring the head to any desired track. Such a machine is termed a moving-head disk drive. An increase in capacity could be obtained by assembling many disks on a common spindle to make a disk pack. The small size of magnetic heads allows the disks to be placed close together. If the positioner is designed so that it can remove the heads away from the disk completely, then it can be exchanged. The exchangeable-pack moving-head disk drive became the standard for mainframe and minicomputers for a long time.

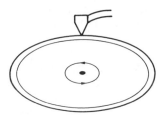

Figure 7.4 The rotating store concept. Data on the rotating circular track are repeatedly presented to the head.

Later came the so-called Winchester technology disks, where the disk and positioner formed a compact sealed unit which allowed increased storage capacity but precluded exchange of the disk pack alone.

Disk drive development has been phenomenally rapid. The first flying-head disks were about three feet across. Subsequently disk sizes of 14, 8, 5¼, 3½ and 1⅞ inches were developed. Despite the reduction in size, the storage capacity is not compromised because the recording density has increased and continues to increase. In fact there is an advantage in making a drive smaller because the moving parts are then lighter and travel a shorter distance, improving access time.

There are numerous types of optical disk, which have different characteristics. There are, however, three broad groups which can be usefully compared:

(1) The Compact Disc and LaserVision are examples of a read-only laser disk, which is designed for mass duplication by stamping. They cannot be recorded.

(2) Some laser disks can be recorded, but once a recording has been made, it cannot be changed or erased. These are usually referred to as write-once-read-many (WORM) disks. The general principle is that the disk contains a thin layer of metal; on recording, a powerful laser melts holes in the layer. Clearly once a pattern of holes has been made, it is permanent.

(3) Erasable optical disks have essentially the same characteristic as magnetic disks, in that new and different recordings can be made in the same track indefinitely, but there is usually a separate erase cycle needed before a new recording can be made since overwrite is not always possible.

Figure 7.5 introduces the essential subsystems of a disk drive which will be discussed here. Magnetic drives and optical drives are similar in that both have a spindle drive mechanism to revolve the disk, and a positioner to give radial access across the disk surface. In the optical drive, the positioner has to carry a collection of lasers, lenses, prisms, gratings and so on, and will be rather larger

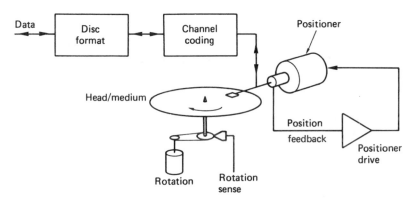

Figure 7.5 The main subsystems of a typical disk drive.

than a magnetic head. The heavier pickup cannot be accelerated as fast as a magnetic drive positioner, and access time is slower. A large number of pickups on one positioner makes matters worse. For this reason and because of the larger spacing needed between the disks, multiplatter optical disks are uncommon. Instead 'juke box' mechanisms have been developed to allow a large library of optical disks to be mechanically accessed by one or more drives. Access time is sometimes reduced by having more than one positioner per disk, a technique adopted rarely in magnetic drives. A penalty of the very small track pitch possible in laser disks, which gives the enormous storage capacity, is that very accurate track following is needed, and it takes some time to lock on to a track. For this reason tracks on laser disks are usually made as a continuous spiral, rather than the concentric rings of magnetic disks. In this way, a continuous data transfer involves no more than track following once the beginning of the file is located.

7.6 Structure of disk

Rigid disks are made from aluminium alloy. Magnetic-oxide types use an aluminium oxide substrate, or undercoat, giving a flat surface to which the oxide binder can adhere. Later metallic disks having higher coercivity are electroplated with the magnetic medium. In both cases the surface finish must be extremely good owing to the very small flying height of the head. As the head-to-disk speed and recording density are functions of track radius, the data are confined to the outer areas of the disks to minimize the change in these parameters. As a result, the centre of the pack is often an empty well. In fixed (i.e. non-interchangeable) disks the drive motor is often installed in the centre well.

The information layer of optical disks may be made of a variety of substances, depending on the working principle. This layer is invariably protected beneath a thick transparent layer of glass or polycarbonate.

Exchangeable optical and magnetic disks are usually fitted in protective cartridges. These have various shutters which retract on insertion in the drive to allow access by the drive spindle and heads. Removable packs usually seat on a taper to ensure concentricity and are held to the spindle by a permanent magnet. A lever mechanism may be incorporated in the cartridge to assist their removal.

7.7 Magnetic disk terminology

In all technologies there are specialist terms, and those relating to magnetic disks will be explained here. Figure 7.6 shows a typical multiplatter magnetic disk pack in conceptual form. Given a particular set of coordinates (cylinder, head, sector), known as a disk physical address, one unique data block is defined. A common block capacity is 512 bytes. The subdivision into sectors is sometimes omitted for special applications. A disk drive can be randomly accessed, because any block address can follow any other, but unlike a RAM, at each address a large block of data is stored, rather than a single word.

7.8 Principle of flying head

Magnetic disk drives permanently sacrifice storage density in order to offer rapid access. The use of a flying head with a deliberate air gap between it and the medium is necessary because of the high medium speed, but this causes a severe separation loss which restricts the linear density available. The air gap must be accurately maintained, and consequently the head is of low mass and is mounted flexibly.

The aerohydrodynamic part of the head is known as the slipper; it is designed to provide lift from the boundary layer which changes rapidly with changes in flying height. It is not initially obvious that the difficulty with disk heads is not making them fly, but making them fly close enough to the disk surface. The boundary layer travelling at the disk surface has the same speed as the disk, but as height increases, it slows down due to drag from the surrounding air. As the lift is a function of relative air speed, the closer the slipper comes to the disk, the greater the lift will be. The slipper is therefore mounted at the end of a rigid cantilever sprung towards the medium. The force with which the head is pressed towards the disk by the spring is equal to the lift at the designed flying height.

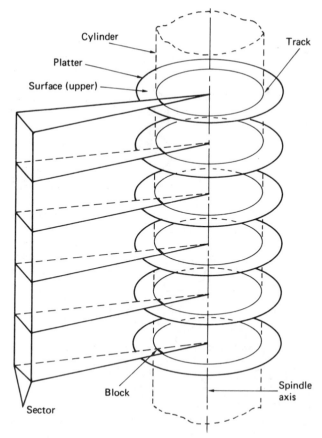

Cylinder

Track

Platter

Surface (upper)

Block

Spindle axis

Sector

Figure 7.6 Disk terminology. Surface: one side of a platter. Track: path described on a surface by a fixed head. Cylinder: imaginary shape intersecting all surfaces at tracks of the same radius. Sector: angular subdivision of pack. Block: that part of a track within one sector. Each block has a unique cylinder, head and sector address.

Because of the spring, the head may rise and fall over small warps in the disk. It would be virtually impossible to manufacture disks flat enough to dispense with this feature. As the slipper negotiates a warp it will pitch and roll in addition to rising and falling, but it must be prevented from yawing, as this would cause an azimuth error. Downthrust is applied to the aerodynamic centre by a spherical thrust button, and the required degrees of freedom are supplied by a thin flexible gimbal. The slipper has to bleed away surplus air in order to approach close enough to the disk, and holes or grooves are usually provided for this purpose in the same way that pinch rollers on some tape decks have grooves to prevent tape slip.

In exchangeable-pack drives, there will be a ramp on the side of the cantilever which engages a fixed block when the heads are retracted in order to lift them away from the disk surface.

Figure 7.7 shows how disk heads are made. The magnetic circuit of disk heads was originally assembled from discrete magnetic elements. As the gap and flying

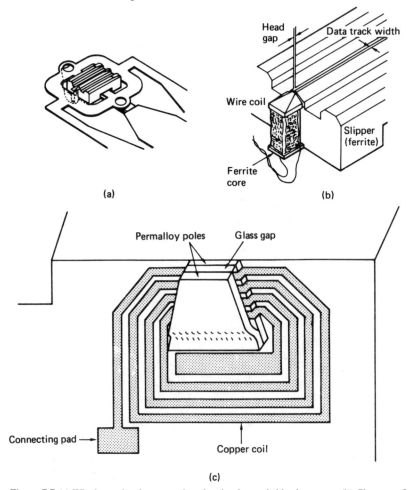

Figure 7.7 (a) Winchester head construction showing large air bleed grooves. (b) Close-up of slipper showing magnetic circuit on trailing edge. (c) Thin film head is fabricated on the end of the slipper using microcircuit technology.

height became smaller to increase linear recording density, the slipper was made from ferrite, and became part of the magnetic circuit. This was completed by a small C-shaped ferrite piece which carried the coil. Ferrite heads were restricted in the coercivity of disk they could write without saturating. In thin-film heads, the magnetic circuit and coil are both formed by deposition on a substrate which becomes the rear of the slipper.

In a moving-head device it is not practicable to position separate erase, record and playback heads accurately. Erase is by overwriting, and reading and writing are carried out by the same head. The presence of the air film causes severe separation loss, and peak shift distortion is a major problem. The flying height of the head varies with the radius of the disk track, and it is difficult to provide accurate equalization of the replay channel because of this. The write current is often controlled as a function of track radius so that the changing reluctance of the air gap does not change the resulting record flux. Automatic gain control (AGC) is used on replay to compensate for changes in signal amplitude from the head.

Equalization may be used on recording in the form of precompensation, which moves recorded transitions in such a way as to oppose the effects of peak shift in addition to any replay equalization used.

Early disks used FM coding, which was easy to decode, but had a poor density ratio. The invention of MFM revolutionized hard disks, and was at one time universal. Further progress led to run-length-limited codes such as 2/3 and 2/7 which had a high density ratio without sacrificing the large jitter window necessary to reject peak shift distortion. Partial response is also suited to disks, but is not yet in common use.

Typical drives have several heads, but with the exception of special-purpose parallel-transfer machines, only one head will be active at any one time, which means that the read and write circuitry can be shared between the heads. Figure 7.8 shows that in one approach the centre-tapped heads are isolated by connecting the centre tap to a negative voltage, which reverse-biases the matrix diodes. The centre tap of the selected head is made positive. When reading, a small current flows through both halves of the head winding, as the diodes are forward-biased. Opposing currents in the head cancel, but read signals due to transitions on the medium can pass through the forward-biased diodes to become differential signals on the matrix bus. During writing, the current from the write generator passes alternately through the two halves of the head coil. Further isolation is necessary to prevent the write-current-induced voltages from destroying the read preamplifier input. Alternatively, FET analog switches may be used for head selection.

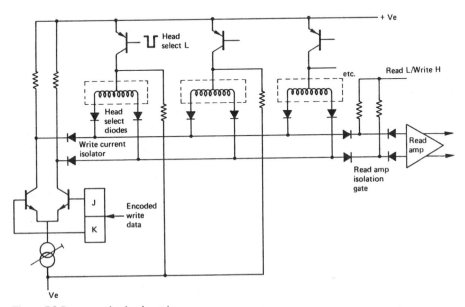

Figure 7.8 Representative head matrix.

The read channel usually incorporates AGC, which will be overridden by the control logic between data blocks in order to search for address marks, which are short unmodulated areas of track. As a block preamble is entered, the AGC will be enabled to allow a rapid gain adjustment.

7.9 Moving the heads

The servo system required to move the heads rapidly between tracks, and yet hold them in place accurately for data transfer, is a fascinating and complex piece of engineering.

In exchangeable-pack drives, the disk positioner moves on a straight axis which passes through the spindle. The head carriage will usually have preloaded ball races which run on rails mounted on the bed of the machine, although some drives use plain sintered bushes sliding on polished rods.

Motive power is generally by moving-coil drive, because of the small moving mass which this technique permits.

When a drive is track-following, it is said to be detented, in fine mode or in linear mode depending on the manufacturer. When a drive is seeking from one track to another, it can be described as being in coarse mode or velocity mode. These are the two major operating modes of the servo.

Moving-coil actuators do not naturally detent and require power to stay on track. The servo system needs positional feedback of some kind. The purpose of the feedback will be one or more of the following:

(1) to count the number of cylinders crossed during a seek
(2) to generate a signal proportional to carriage velocity
(3) to generate a position error proportional to the distance from the centre of the desired track.

Magnetic and optical drives obtain these feedback signals in different ways. Many drives incorporate a tacho which may be a magnetic moving-coil type or

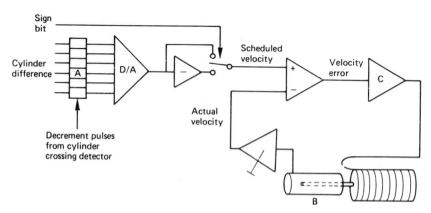

Figure 7.9 Control of carriage velocity by cylinder difference. The cylinder difference is loaded into the difference counter A. A digital-to-analog converter generates an analog voltage from the cylinder difference, known as the scheduled velocity. This is compared with the actual velocity from the transducer B in order to generate the velocity error which drives the servo amplifier C.

its complementary equivalent, the moving-magnet type. Both generate a voltage proportional to velocity, and can give no positional information.

A seek is a process where the positioner moves from one cylinder to another. The speed with which a seek can be completed is a major factor in determining the access time of the drive. The main parameter controlling the carriage during a seek is the cylinder difference, which is obtained by subtracting the current cylinder address from the desired cylinder address. The cylinder difference will be a signed binary number representing the number of cylinders to be crossed to reach the target, direction being indicated by the sign. The cylinder difference is loaded into a counter which is decremented each time a cylinder is crossed. The counter drives a DAC which generates an analog voltage proportional to the cylinder difference. As Figure 7.9 shows, this voltage, known as the scheduled velocity, is compared with the output of the carriage-velocity tacho. Any difference between the two results in a velocity error which drives the carriage to cancel the error. As the carriage approaches the target cylinder, the cylinder difference becomes smaller, with the result that the run-in to the target is critically damped to eliminate overshoot.

Figure 7.10(a) shows graphs of scheduled velocity, actual velocity and motor current with respect to cylinder difference during a seek. In the first half of the seek, the actual velocity is less than the scheduled velocity, causing a large velocity error which saturates the amplifier and provides maximum carriage acceleration. In the second half of the graphs, the scheduled velocity is falling below the actual velocity, generating a negative velocity error which drives a reverse current through the motor to slow the carriage down. The scheduled deceleration slope can clearly not be steeper than the saturated acceleration slope.

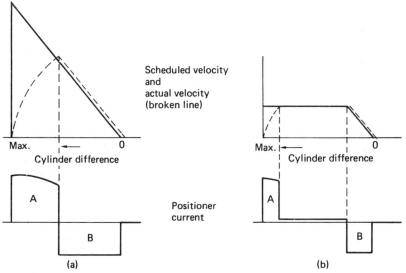

Figure 7.10 In the simple arrangement in (a) the dissipation in the positioner is continuous, causing a heating problem. The effect of limiting the scheduled velocity above a certain cylinder difference is apparent in (b) where heavy positioner current only flows during acceleration and deceleration. During the plateau of the velocity profile, only enough current to overcome friction is necessary. The curvature of the acceleration slope is due to the back EMF of the positioner motor.

Areas A and B on the graph will be about equal, as the kinetic energy put into the carriage has to be taken out. The current through the motor is continuous, and would result in a heating problem, so to counter this, the DAC is made non-linear so that above a certain cylinder difference no increase in scheduled velocity will occur. This results in the graph of Figure 7.10(b). The actual velocity graph is called a velocity profile. It consists of three regions: acceleration, where the system is saturated; a constant velocity plateau, where the only power needed is to overcome friction; and the scheduled run-in to the desired cylinder. Dissipation is only significant in the first and last regions.

7.10 Rotation

The rotation subsystems of disk drives will now be covered. The track-following accuracy of a drive positioner will be impaired if there is bearing run-out, and so the spindle bearings are made to a high degree of precision. Most modern drives incorporate brushless DC motors with integral speed control. In exchangeable-pack drives, some form of braking is usually provided to slow down the pack for convenient removal.

In order to control reading and writing, the drive control circuitry needs to know which cylinder the heads are on, and which sector is currently under the head. Sector information used to be obtained from a sensor which detects holes or slots cut in the hub of the disk. Modern drives will obtain this information from the disk surface as will be seen. The result is that a sector counter in the control logic remains in step with the physical rotation of the disk. The desired sector address is loaded into a register, which is compared with the sector counter. When the two match, the desired sector has been found. This process is referred to as a search, and usually takes place after a seek. Having found the correct physical place on the disk, the next step is to read the header associated with the data block to confirm that the disk address contained there is the same as the desired address.

7.11 Servo-surface disks

One of the major problems to be overcome in the development of high-density disk drives was that of keeping the heads on track despite changes of temperature. The very narrow tracks used in digital recording have similar dimensions to the amount a disk will expand as it warms up. The cantilevers and the drive base all expand and contract, conspiring with thermal drift in the cylinder transducer to limit track pitch. The breakthrough in disk density came with the introduction of the servo-surface drive. The position error in a servo-surface drive is derived from a head reading the disk itself. This virtually eliminates thermal effects on head positioning and allows great increases in storage density.

In a multiplatter drive, one surface of the pack holds servo information which is read by the servo head. In a ten-platter pack this means that 5% of the medium area is lost, but this is unimportant since the increase in density allowed is enormous. Using one side of a single-platter cartridge for servo information would be unacceptable as it represents 50% of the medium area, so in this case

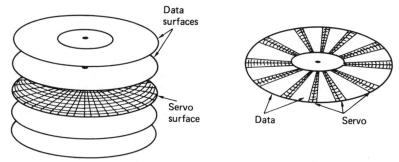

Data
surfaces

Servo
surface

Data

Servo

Figure 7.11 In a multiplatter disk pack, one surface is dedicated to servo information. In a single platter, the servo information is embedded in the data on the same surfaces.

the servo information can be interleaved with sectors on the data surfaces. This is known as an embedded-servo technique. These two approaches are contrasted in Figure 7.11.

The servo surface is written at the time of disk pack manufacture, and the disk drive can only read it. Writing the servo surface has nothing to do with disk formatting, which affects the data storage areas only.

As there are exactly the same number of pulses on every track on the servo surface, it is possible to describe the rotational position of the disk simply by counting them. All that is needed is a unique pattern of missing pulses once per revolution to act as an index point, and the sector transducer can also be eliminated.

The advantage of deriving the sector count from the servo surface is that the number of sectors on the disk can be varied. Any number of sectors can be accommodated by feeding the pulse signal through a programmable divider, so the same disk and drive can be used in numerous different applications.

7.12 Winchester technology

In order to offer extremely high capacity per spindle, which reduces the cost per bit, a disk drive must have very narrow tracks placed close together, and must use very short recorded wavelengths, which implies that the flying height of the heads must be small. The so-called Winchester technology is one approach to high storage density. The technology was developed by IBM, and the name came about because the model number of the development drive was the same as that of the famous rifle.

Reduction in flying height magnifies the problem of providing a contaminant-free environment. A conventional disk is well protected whilst inside the drive, but outside the drive the effects of contamination become intolerable.

In exchangeable-pack drives, there is a real limit to the track pitch that can be achieved because of the difficulty or cost of engineering head-alignment mechanisms to make the necessary minute adjustments to give interchange compatibility.

The essence of Winchester technology is that each disk pack has its own set of read/write and servo heads, with an integral positioner. The whole is protected by a dust-free enclosure, and the unit is referred to as a head disk assembly, or HDA.

As the HDA contains its own heads, compatibility problems do not exist, and no head alignment is necessary or provided for. It is thus possible to reduce track pitch considerably compared with exchangeable-pack drives. The sealed environment ensures complete cleanliness which permits a reduction in flying height without loss of reliability, and hence leads to an increased linear density. If the rotational speed is maintained, this can also result in an increase in data transfer rate.

The HDA is completely sealed, but some have a small filtered port to equalize pressure. Into this sealed volume of air, the drive motor delivers the majority of its power output. The resulting heat is dissipated by fins on the HDA casing. Some HDAs are filled with helium which significantly reduces drag and heat build-up.

An exchangeable-pack drive must retract the heads to facilitate pack removal. With Winchester technology this is not necessary. An area of the disk surface is reserved as a landing strip for the heads. The disk surface is lubricated, and the heads are designed to withstand landing and take-off without damage. Winchester heads have very large air-bleed grooves to allow low flying height with a much smaller downthrust from the cantilever, and so they exert less force on the disk surface during contact. When the term *parking* is used in the context of Winchester technology, it refers to the positioning of the heads over the landing area.

Disk rotation must be started and stopped quickly to minimize the length of time the heads slide over the medium. A powerful motor will accelerate the pack quickly. Eddy-current braking cannot be used, since a power failure would allow the unbraked disk to stop only after a prolonged head contact period. A failsafe

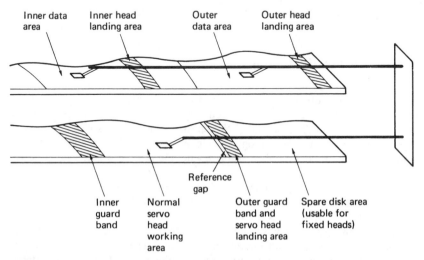

Figure 7.12 When more than one head is used per surface, the positioner still only requires one servo head. This is often arranged to be equidistant from the read/write heads for thermal stability.

mechanical brake is used, which is applied by a spring and released with a solenoid.

A major advantage of contact start/stop is that more than one head can be used on each surface if retraction is not needed. This leads to two gains: first, the travel of the positioner is reduced in proportion to the number of heads per surface, reducing access time; and, second, more data can be transferred at a given detented carriage position before a seek to the next cylinder becomes necessary. This increases the speed of long transfers. Figure 7.12 illustrates the relationships of the heads in such a system.

7.13 Rotary positioners

Figure 7.13 shows that rotary positioners are feasible in Winchester drives; they cannot be used in exchangeable-pack drives because of interchange problems. There are some advantages to a rotary positioner. It can be placed in the corner of a compact HDA allowing smaller overall size. The manufacturing cost will be less than a linear positioner because fewer bearings and precision bars are

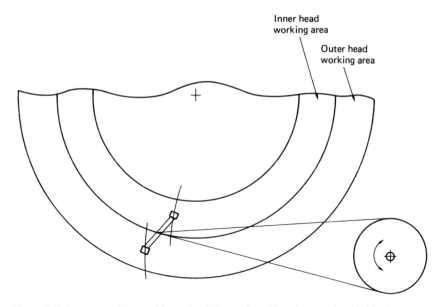

Figure 7.13 A rotary positioner with two heads per surface. The tolerances involved in the spacing between the heads and the axis of rotation mean that each arm records data in a unique position. Those data can only be read back by the same heads, which rules out the use of a rotary positioner in exchangeable-pack drives. In a head disk assembly the problem of compatibility does not arise.

needed. Significantly, a rotary positioner can be made faster since its inertia is smaller. With a linear positioner all parts move at the same speed. In a rotary positioner, only the heads move at full speed, as the parts closer to the shaft must move more slowly. Figure 7.14 shows a typical HDA with a rotary positioner. The principle of many rotary positioners is exactly that of a moving-coil ammeter, where current is converted directly into torque.

Breather filter

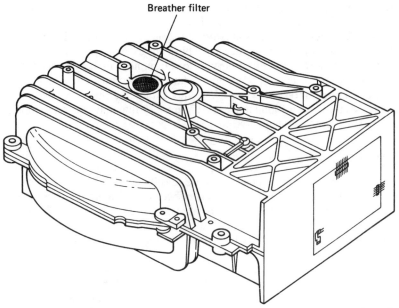

Figure 7.14 Head disk assembly with a rotary positioner. The adoption of this technique allows a very compact structure.

One disadvantage of rotary positioners is that there is a component of windage on the heads which tends to pull the positioner in towards the spindle. In linear positioners windage is at right angles to motion and can be neglected. Windage can be overcome in rotary positioners by feeding the current cylinder address to a ROM which sends a code to a DAC. This produces an offset voltage which is fed to the positioner driver to generate a torque which balances the windage whatever the position of the heads.

When extremely small track spacing is contemplated, it cannot be assumed that all the heads will track the servo head due to temperature gradients. In this case the embedded-servo approach must be used, where each head has its own alignment patterns. The servo surface is often retained in such drives to allow coarse positioning, velocity feedback and index and write clock generation, in addition to locating the guard bands for landing the heads.

Winchester drives have been made with massive capacity, but the problem of backup is then magnified, and the general trend has been for the physical size of the drive to come down as the storage density increases in order to improve access time. Very small Winchester disk drives are now available which plug into standard integrated circuit sockets. These are competing with RAM for memory applications where non-volatility is important.

7.14 Optical disk principles

In order to record MO disks or replay any optical disk, a source of monochromatic light is required. The light source must have low noise otherwise the variations in intensity due to the noise of the source will mask the variations due to reading the disk. The requirement for a low-noise monochromatic light source is economically met using a semiconductor laser.

In the LED, the light produced is incoherent or noisy. In the laser, the ends of the semiconductor are optically flat mirrors, which produce an optically resonant cavity. One photon can bounce to and fro, exciting others in synchronism, to produce coherent light. This is known as light amplification by stimulated emission of radiation, mercifully abbreviated to LASER, and can result in a runaway condition, where all available energy is used up in one flash. In injection lasers, an equilibrium is reached between energy input and light output, allowing continuous operation with a clean output. The equilibrium is delicate, and such devices are usually fed from a current source. To avoid runaway when temperature change disturbs the equilibrium, a photosensor is often fed back to the current source. Such lasers have a finite life, and become steadily less efficient. The feedback will maintain output, and it is possible to anticipate the failure of the laser by monitoring the drive voltage needed to give the correct output.

Re-recordable or erasable optical disks rely on magneto-optics, also known more fully as thermomagneto-optics. Writing in such a device makes use of a thermomagnetic property posessed by all magnetic materials, which is that above a certain temperature, known as the Curie temperature, their coercive force becomes zero. This means that they become magnetically very soft, and take on the flux direction of any externally applied field. On cooling, this field orientation will be frozen in the material, and the coercivity will oppose attempts to change it. Although many materials possess this property, there are relatively few which have a suitably low Curie temperature. Compounds of terbium and gadolinium have been used, and one of the major problems to be overcome is that almost all suitable materials from a magnetic viewpoint corrode very quickly in air.

Figure 7.15 shows how most magneto-optical disks are written. The intensity of the laser is modulated with the waveform to be recorded. If the disk is considered to be initially magnetized along its axis of rotation with the north pole upwards, it is rotated in a field of the opposite sense, produced by a steady

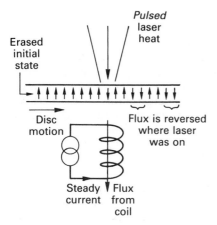

Figure 7.15 MO discs can be written by modulating the laser.

current flowing in a coil which is weaker than the room-temperature coercivity of the medium. The field will therefore have no effect. A pulse from the laser will momentarily heat a very small area of the medium past its Curie temperature, whereby it will take on a reversed flux due to the presence of the field coils. This reversed-flux direction will be retained indefinitely as the medium cools.

The storage medium is clearly magnetic, but the writing mechanism is the heat produced by light from a laser; hence the term thermomagneto-optics. The advantage of this writing mechanism is that there is no physical contact between the writing head and the medium. The distance can be several millimetres, some of which is taken up with a protective layer to prevent corrosion. In prototypes, this layer is glass, but commercially available disks use plastics.

The laser beam will supply a relatively high power for writing, since it is supplying heat energy. For reading, the laser power is reduced, such that it cannot heat the medium past the Curie temperature, and it is left on continuously. Readout depends on the so-called Kerr effect, which describes a rotation of the plane of polarization of light due to a magnetic field. The magnetic areas written on the disk will rotate the plane of polarization of incident polarized light to two different planes, and it is possible to detect the change in rotation with a suitable pickup.

7.15 Optical pickups

Whatever the type of disk being read, it must be illuminated by the laser beam. Some of the light reflected back from the disk re-enters the aperture of the objective lens. The pickup must be capable of separating the reflected light from the incident light. When playing MO disks, the intensity does not change, but the magnetic recording on the disk rotates the plane of polarization one way or the other depending on the direction of the vertical magnetization. Figure 7.16(a) shows that a polarizing prism is required to linearly polarize the light from the laser on its way to the disk. Light returning from the disk has had its plane of polarization rotated by approximately ±1 degree. This is an extremely small rotation. Figure 7.16(b) shows that the returning rotated light can be considered to be comprised of two orthogonal components. R_x is the component which is in the same plane as the illumination and is called the *ordinary* component and R_y is the component due to the Kerr effect rotation and is known as the *magneto-optic* component. A polarizing beam splitter mounted squarely would reflect the magneto-optic component R_y very well because it is at right angles to the transmission plane of the prism, but the ordinary component would pass straight on in the direction of the laser. By rotating the prism slightly a small amount of the ordinary component is also reflected. Figure 7.16(c) shows that when combined with the magneto-optic component, the angle of rotation has increased. Detecting this rotation requires a further polarizing prism or analyser as shown in Figure 7.16(a). The prism is twisted such that the transmission plane is at 45 degrees to the planes of R_x and R_y. Thus with an unmagnetized disk, half of the light is transmitted by the prism and half is reflected. If the magnetic field of the disk turns the plane of polarization towards the transmission plane of the prism, more light is transmitted and less is reflected. Conversely if the plane of polarization is rotated away from the transmission plane, less light is transmitted and more is reflected. If two sensors are used, one for transmitted light and one for reflected light, the difference between the two sensor outputs will be a

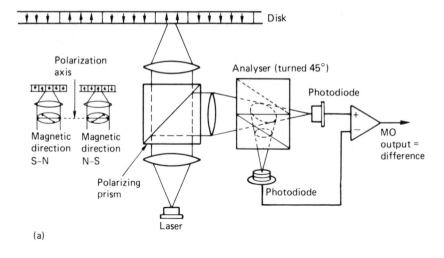

(a)

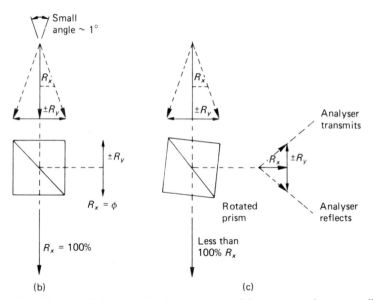

(b)   (c)

Figure 7.16 A pickup suitable for the replay of magneto-optic disks must respond to very small rotations of the plane of polarization.

waveform representing the angle of polarization and thus the recording on the disk. This differential analyser eliminates common mode noise in the reflected beam.

High-density recording implies short wavelengths. Using a laser focused on the disk from a distance allows short-wavelength recordings to be played back without physical contact, whereas conventional magnetic recording requires intimate contact and implies a wear mechanism, the need for periodic cleaning, and susceptibility to contamination.

The information layer is read through the thickness of the disk; this approach causes the readout beam to enter and leave the disk surface through the largest possible area. Despite the minute spot size of about 1 micrometre diameter, light enters and leaves through a 1 mm diameter circle. As a result, surface debris has to be three orders of magnitude larger than the readout spot before the beam is obscured. This approach has the further advantage in MO drives that the magnetic head, on the opposite side to the laser pickup, is then closer to the magnetic layer in the disk.

7.16 The disk controller

A disk controller is a unit which is interposed between the drives and the rest of the system. It consists of two main parts: that which issues control signals to and obtains status from the drives, and that which handles the data to be stored and retrieved. Both parts are synchronized by the control sequencer. The essentials of a disk controller are determined by the characteristics of drives and the functions needed, and so they do not vary greatly. It is desirable for economic reasons to use a commercially available disk controller intended for computers. Such controllers are adequate for still store applications, but cannot support the data rate required for real time moving video unless data reduction is employed. Disk drives are generally built to interface to a standard controller interface, such as the SCSI bus. The disk controller will then be a unit which interfaces the drive bus to the host computer system.

The execution of a function by a disk subsystem requires a complex series of steps, and decisions must be made between the steps to decide what the next will be. There is a parallel with computation, where the function is the equivalent of an instruction, and the sequencer steps needed are the equivalent of the microinstructions needed to execute the instruction. The major failing in this analogy is that the sequence in a disk drive must be accurately synchronized to the rotation of the disk.

Most disk controllers use direct memory access, which means that they have the ability to transfer disk data in and out of the associated memory without the assistance of the processor. In order to cause a file transfer, the disk controller must be told the physical disk address (cylinder, sector, track), the physical memory address where the file begins, the size of the file and the direction of transfer (read or write). The controller will then position the disk heads, address the memory, and transfer the samples. One disk transfer may consist of many contiguous disk blocks, and the controller will automatically increment the disk-address registers as each block is completed. As the disk turns, the sector address increases until the end of the track is reached. The track or head address will then be incremented and the sector address reset so that transfer continues at the beginning of the next track. This process continues until all of the heads have been used in turn. In this case both the head address and sector address will be reset, and the cylinder address will be incremented, which causes a seek. A seek which takes place because of a data transfer is called an implied seek, because it is not necessary formally to instruct the system to perform it. As disk drives are block-structured devices, and the error correction is codeword based, the controller will always complete a block even if the size of the file is less than a whole number of blocks. This is done by packing the last block with zeros.

The status system allows the controller to find out about the operation of the drive, both as a feedback mechanism for the control process, and to handle any errors. Upon completion of a function, it is the status system which interrupts the control processor to tell it that another function can be undertaken.

In a system where there are several drives connected to the controller via a common bus, it is possible for non-data-transfer functions such as seeks to take place in some drives simultaneously with a data transfer in another.

Before a data transfer can take place, the selected drive must physically access the desired block, and confirm this by reading the block header. Following a seek to the required cylinder, the positioner will confirm that the heads are on track and settled. The desired head will be selected, and then a search for the correct sector begins. This is done by comparing the desired sector with the current sector register, which is typically incremented by dividing down servo-surface pulses. When the two counts are equal, the head is about to enter the desired block. Figure 7.17 shows the structure of a typical magnetic disk track. In between blocks are placed address marks, which are areas without transitions which the read circuits can detect. Following detection of the address mark, the sequencer is roughly synchronized to begin handling the block. As the block is entered, the data separator locks to the preamble, and in due course the sync pattern will be found. This sets to zero a counter which divides the data bit rate by eight, allowing the serial recording to be correctly assembled into bytes, and also allowing the sequencer to count the position of the head through the block in order to perform all the necessary steps at the right time.

The first header word is usually the cylinder address, and this is compared with the contents of the desired cylinder register. The second header word will contain the sector and track address of the block, and these will also be compared with the desired addresses. There may also be bad-block flags and/or defect-skipping information. At the end of the header is a CRCC which will be used to ensure that

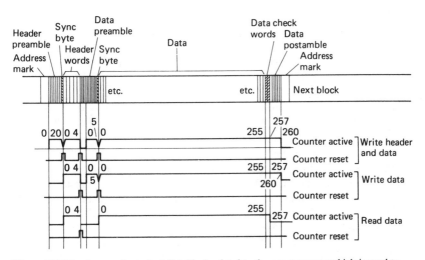

Figure 7.17 The format of a typical disk block related to the count process which is used to establish where in the block the head is at any time. During a read the count is derived from the actual data read, but during a write, the count is derived from the write clock.

the header was read correctly. Figure 7.18 shows a flowchart of the position verification, after which a data transfer can proceed. The header reading is completely automatic. The only time it is necessary formally to command a header to be read is when checking that a disk has been formatted correctly.

During the read of a data block, the sequencer is employed again. The sync pattern at the beginning of the data is detected as before, following which the actual data arrive. These bits are converted to byte or sample parallel, and sent to the memory by DMA. When the sequencer has counted the last data byte off the track, the redundancy for the error-correction system will be following.

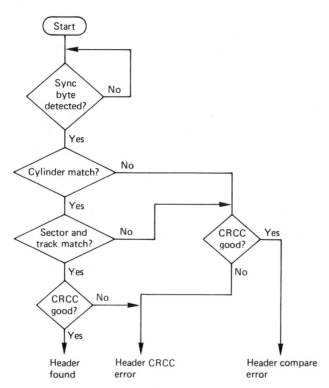

Figure 7.18 The vital process of position confirmation is carried out in accordance with the above flowchart. The appropriate words from the header are compared in turn with the contents of the disk-address registers in the subsystem. Only if the correct header has been found and read properly will the data transfer take place.

During a write function, the header-check function will also take place as it is perhaps even more important not to write in the wrong place on a disk. Once the header has been checked and found to be correct, the write process for the associated data block can begin. The preambles, sync pattern, data block, redundancy and postamble have all to be written contiguously. This is taken care of by the sequencer, which is obtaining timing information from the servo surface to lock the block structure to the angular position of the disk. This should be contrasted with the read function, where the timing comes directly from the data.

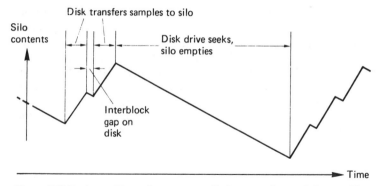

Figure 7.19 During a video replay sequence, silo is constantly emptied to provide samples, and is refilled in blocks by the drive.

When video samples are fed into a disk-based system, from a digital interface or from an ADC, they will be placed in a memory, from which the disk controller will read them by DMA. The continuous input sample stream will be split up into disk blocks for disk storage.

The disk transfers must by definition be intermittent, because there are headers between contiguous sectors. Once all the sectors on a particular cylinder have been used, it will be necessary to seek to the next cylinder, which will cause a further interruption to the data transfer. If a bad block is encountered, the sequence will be interrupted until it has passed. The instantaneous data rate of a parallel transfer drive is made higher than the continuous video data rate, so that there is time for the positioner to move whilst the video output is supplied from the FIFO memory. In replay, the drive controller attempts to keep the FIFO as full as possible by issuing a read command as soon as one block space appears in the FIFO. This allows the maximum time for a seek to take place before reading must resume. Figure 7.19 shows the action of the FIFO. Whilst recording, the drive controller attempts to keep the FIFO as empty as possible by issuing write commands as soon as a block of data is present. In this way the amount of time available to seek is maximized in the presence of a continuous video sample input.

7.17 Defect handling

The protection of data recorded on disks differs considerably from the approach used on other media in digital video. This has much to do with the intolerance of data processors to errors when compared with video data. In particular, it is not possible to interpolate to conceal errors in a computer program or a data file.

In the same way that magnetic tape is subject to dropouts, magnetic disks suffer from surface defects whose effect is to corrupt data. The shorter wavelengths employed as disk densities increase are affected more by a given size of defect. Attempting to make a perfect disk is subject to a law of diminishing returns, and eventually a state is reached where it becomes more cost effective to invest in a defect-handling system.

In the construction of bad-block files, a brand new disk is tested by the operating system. Known patterns are written everywhere on the disk, and these are read back and verified. Following this the system gives the disk a volume name, and creates on it a directory structure which keeps records of the position and size of every file subsequently written. The physical disk address of every block which fails to verify is allocated to a file which has an entry in the disk directory. In this way, when genuine data files come to be written, the bad blocks appear to the system to be in use storing a fictitious file, and no attempt will be made to write there. Some disks have dedicated tracks where defect information can be written during manufacture or by subsequent verification programs, and these permit a speedy construction of the system bad-block file.

Glossary

Accumulator Logic circuit which adds a series of numbers which are fed to it.

AES/EBU interface Standardized interface for transmitting digital audio information between two devices (*see* Channel Status).

ATV Advanced television; television transmission systems using modern techniques such as compression and error correction.

Aliasing Beat frequencies produced when sampling rate (q.v.) is not high enough.

Aspect ratio Ratio of image width to height. SDTV is 4:3; widescreen TV is 16:9.

Azimuth recording Magnetic recording technique which reduces crosstalk between adjacent tracks.

BER Bit Error Rate (*see* Bit).

Bit Abbreviation for Binary Digit.

Block matching Simple technique used to estimate motion between images in compression and standards conversion.

BNC Bayonet Neill-Concelman; coaxial connector used for video signals.

Byte Group or word of bits (q.v.) generally eight.

Channel coding Method of expressing data as a waveform which can be recorded or transmitted.

Codeword Entity used in error correction which has constant testable characteristic.

Coefficient Pretentious word for a binary number used to control a multiplier.

Coercivity Measure of the erasure difficulty, hence replay energy, of a magnetic recording.

Companding Abbreviation of compressing and expanding; means for increasing dynamic range.

Concealment Method of making uncorrectable errors less visible; e.g. interpolation.

Contouring Video artefact caused by insufficient sample wordlength.

Crosstalk Unwanted signal breaking through from adjacent wiring or recording track.

Curie temperature Temperature at which magnetic materials demagnetize.

Cylinder In disks, set of tracks having same radius.

DCT   Discrete Cosine Transform; transform used in compression based on rectangular pixel blocks.

Decimation   Reduction of sampling rate by omitting samples.

Dither   Noise added to analog signal to linearize quantizer.

Downconvertor   Unit which allows HDTV signals to be converted to standard definition.

DSP   Digital Signal Processor; computer optimized for waveform processing.

EDH   Error Detection and Handling; a data integrity option for SDI digital interface (q.v.).

EDL   Edit Decision List; used to control editing process with timecode.

EDTV   Extended Definition Television; television systems which enhance the quality of existing TV standards.

EFM   Eight to Fourteen Modulation; the channel code (q.v.) used in D-3 and D-5 DVTRs.

EMC   Electromagnetic Compatibility; legislation controlling sensitivity of equipment to external interference and limiting radiation of interference.

Embedded audio   Digital audio signals multiplexed into ancillary data capacity of SDI (q.v.).

Entropy   The useful information in a signal.

Eye pattern   Characteristic pattern seen on oscilloscope when viewing channel coded (q.v.) signals.

Faraday effect   Rotation of plane of polarization of light by magnetic field.

Ferrite   Hard non-conductive magnetic material used for heads and transformers.

Flash convertor   High speed ADC used for video conversion.

Fourier transform   Frequency domain or spectral representation of a signal.

Galois field   Mathematical entity on which Reed–Solomon (q.v.) error correction is based.

HDTV   High Definition Television; television systems which have a large number of scanning lines.

Hamming distance   Number of bits different between two words.

Interfield compression   Compression technique which exploits similarities between successive fields (*see* MPEG).

Interleaving   Reordering of data on medium to reduce effect of defects.

Interpolation   Replacing missing sample with average of those either side.

Intrafield compression   Compression technique which considers video fields in isolation from their neighbours so that e.g. editing is possible (*see* JPEG).

JPEG   Joint Photographic Experts Group; ISO standard for still image or single frame compression.

Jitter   Digital equivalent of timebase error.

Kerr effect   *see* Faraday effect.

MTF   Modulation transfer function, measure of the resolving power of optical system.

MPEG   Moving Picture Experts Group; ISO standard for moving image compression.

Motion compensation   Technique used to eliminate judder in standards convertor (*see* Phase correlation; Block matching; Motion vector).

Motion vector   Parameter describing direction and speed of motion of part of image.

Non-linear Video editing technique using random access storage such as disk drives.

Oversampling Use of sampling rate which is higher than necessary.

Partial response Channel coding (q.v.) technique used in Digital Betacam.

Phase correlation Technique used for motion estimation in compression and standards conversion.

Phase Linear Describes a circuit which has constant delay at all frequencies.

Pixel Picture cell or element; one sample of image.

Product Code Combination of two one-dimensional error correcting codes in an array.

Pseudo-random code Number sequence which is sufficiently random for practical purposes but which is repeatable.

Reconstruction Creating continuous analog signal from samples.

Reed–Solomon code Error correction code which is popular because it is as powerful as theory allows.

Requantizing Shortening of sample wordlength.

SAV Start of Active Video; (*see* TRS).

SDI Serial Digital Interface; standardized coaxial cable digital video interface.

Sampling rate Frequency with which samples are taken (*see* Aliasing).

Seek Moving the heads on a disk drive.

Shuffle Irregular reordering of pixels to ease concealment (q.v) of uncorrectable errors.

Signature analysis Test technique for verifying quality of digital transmission system.

Spline Motion interpolating algorithm used in DVEs.

Square Pixels Term used to indicate that the spacing between pixels is the same horizontally and vertically.

TRS Timing Reference Signal; bit pattern used in digital video interfaces for synchronizing.

Upconvertor Device which allows standard definition signals to be displayed on HDTV equipment.

Wordlength Number of bits in a sample; typically 8 or 10 in video.

Index